International Conference EcoBalt 2023
"Chemicals & Environment"

International Conference EcoBalt 2023 "Chemicals & Environment"

Editors

Monika Mortimer
Anne Kahru
Ivo Leito
Riin Rebane
Villem Aruoja

MDPI

Basel • Beijing • Wuhan • Barcelona • Belgrade • Novi Sad • Cluj • Manchester

Editors
Monika Mortimer
National Institute of Chemical Physics and Biophysics
Tallinn
Estonia

Anne Kahru
National Institute of Chemical Physics and Biophysics
Tallinn
Estonia

Ivo Leito
University of Tartu
Tartu
Estonia

Riin Rebane
University of Tartu
Tartu
Estonia

Villem Aruoja
National Institute of Chemical Physics and Biophysics
Tallinn
Estonia

Editorial Office
MDPI AG
Grosspeteranlage 5
4052 Basel, Switzerland

This is a reprint of articles from the Proceedings published online in the open access journal *Proceedings* (ISSN 2504-3900) (available at: https://www.mdpi.com/2504-3900/92/1).

For citation purposes, cite each article independently as indicated on the article page online and as indicated below:

Lastname, A.A.; Lastname, B.B. Article Title. *Journal Name* **Year**, *Volume Number*, Page Range.

ISBN 978-3-7258-1557-9 (Hbk)
ISBN 978-3-7258-1558-6 (PDF)
doi.org/10.3390/books978-3-7258-1558-6

Contents

proceedings

Editorial

Statement of Peer Review †

Monika Mortimer [1,*], Anne Kahru [1], Ivo Leito [2], Riin Rebane [2,3] and Villem Aruoja [1]

1 Laboratory of Environmental Toxicology, National Institute of Chemical Physics and Biophysics, Akadeemia tee 23, 12618 Tallinn, Estonia; anne.kahru@kbfi.ee (A.K.); villem.aruoja@kbfi.ee (V.A.)
2 Institute of Chemistry, University of Tartu, Ravila 14a, 50411 Tartu, Estonia; ivo.leito@ut.ee (I.L.); riin.rebane@klab.ee (R.R.)
3 Estonian Environmental Research Centre, Marja 4d, 10617 Tallinn, Estonia
* Correspondence: monika.mortimer@kbfi.ee
† All papers published in the volume are presented at the International Conference EcoBalt 2023 "Chemicals & Environment", Tallinn, Estonia, 9–11 October 2023.

In submitting conference proceedings to *Proceedings*, the volume editors of the proceedings certify to the publisher that all papers published in this volume have been subjected to peer review administered by the volume editors. Reviews were conducted by expert referees to the professional and scientific standards expected of a *Proceedings* journal.

- Type of peer review: single-blind.
- Conference submission management system: Google Forms.
- Number of submissions sent for review: 104.
- Number of submissions accepted: 104.
- Acceptance rate (number of submissions accepted/number of submissions received): 100%.
- Average number of reviews per paper: 1.
- Total number of reviewers involved: 5.
- Any additional information on the review process: N/A.

Conflicts of Interest: The authors declare no conflict of interest.

Disclaimer/Publisher's Note: The statements, opinions and data contained in all publications are solely those of the individual author(s) and contributor(s) and not of MDPI and/or the editor(s). MDPI and/or the editor(s) disclaim responsibility for any injury to people or property resulting from any ideas, methods, instructions or products referred to in the content.

Citation: Mortimer, M.; Kahru, A.; Leito, I.; Rebane, R.; Aruoja, V. Statement of Peer Review. *Proceedings* **2023**, *92*, 3. https://doi.org/10.3390/proceedings2023092003

Published: 20 November 2023

proceedings

MDPI

Editorial

The 23rd Biennial Conference EcoBalt 2023 in Tallinn, Estonia †

Monika Mortimer [1,*], Anne Kahru [1], Ivo Leito [2], Riin Rebane [2,3] and Villem Aruoja [1]

1 Laboratory of Environmental Toxicology, National Institute of Chemical Physics and Biophysics, Akadeemia Tee 23, 12618 Tallinn, Estonia; anne.kahru@kbfi.ee (A.K.); villem.aruoja@kbfi.ee (V.A.)
2 Institute of Chemistry, University of Tartu, Ravila 14a, 50411 Tartu, Estonia; ivo.leito@ut.ee (I.L.); riin.rebane@klab.ee (R.R.)
3 Estonian Environmental Research Centre, Marja 4d, 10617 Tallinn, Estonia
* Correspondence: monika.mortimer@kbfi.ee
† All papers published in the volume are presented at the International Conference EcoBalt 2023 "Chemicals & Environment", Tallinn, Estonia, 9–11 October 2023.

The 23rd biennial conference EcoBalt2023 took place in Tallinn, Estonia, continuing the tradition of international EcoBalt conferences in bringing together scientists, students, educators, regulators, and industry representatives from the Baltic States and nearby regions.

EcoBalt2023 served as a forum for environmental scientists, ecologists, analytical and organic chemists, material scientists, toxicologists, and risk assessors to discuss current topics related to the environment.

The conference topics included the following:

- Environmental chemistry
- Environmental toxicology
- Ecology
- Analytical chemistry
- Green and sustainable chemistry
- Implications and applications of nanomaterials

During two and a half days of the on-site event, nearly 200 participants from 21 countries (Figure 1) had a chance to listen to 59 talks from experts in their fields, as well as from early career researchers. In addition, 45 posters were presented during the conference.

The conference started with a welcome event and a special lecture by Dr. Priit Zingel from the Estonian University of Life Sciences, Tartu, Estonia. Dr. Zingel presented a captivating talk titled "The Challenge of Understanding—from Protozoa to Ecosystems", encouraging the audience to ponder on how the surrounding nature exchanges, collects, and manages information—both at the ecosystem and cellular levels.

The first day of EcoBalt 2023 started off with a presentation by the first plenary speaker—Dr. Stephen Ellison—a Science Fellow at LGC, the UK National Measurement Laboratory for chemical and biological measurement. Dr. Ellison is an expert in statistics, measurement uncertainty, and reference material certification who has co-authored EURACHEM guides on measurement uncertainty, metrological traceability, and qualitative analysis. Dr. Ellison's talk was titled "Sampling, Detection and Uncertainty in Environmental Analysis—Challenges and Solutions". The plenary lecture was followed by parallel sessions of talks by experts in the field as well as early career researchers. The topics of the talks included data quality in environmental analysis, the effects of legacy and emerging chemicals (such as microplastics) on ecosystems and humans, and the water quality of Narva River, Lake Peipsi, and the Baltic Sea. The topic of micro- and nanoplastic effects on the environment and human health was also covered by Prof. Pu Chun Ke, the plenary speaker on the second day of EcoBalt 2023. Prof. Ke from the Nanomedicine Center of The Great Bay Area National Institute for Nanotechnology Innovation (Guangzhou, China) and Monash University (Melbourne, Australia) is an expert in nanomaterial–biomolecular interactions and his multidisciplinary research career has spanned over three continents

Citation: Mortimer, M.; Kahru, A.; Leito, I.; Rebane, R.; Aruoja, V. The 23rd Biennial Conference EcoBalt 2023 in Tallinn, Estonia. *Proceedings* **2023**, *92*, 26. https://doi.org/10.3390/proceedings2023092026

Published: 21 November 2023

and has been funded by NSF and EPA in the United States, CSIRO and CBNS in Australia, and NSFC and MOST in China.

Figure 1. Participants of EcoBalt 2023.

The conference included special sessions organized and funded by the Estonian Society of Toxicology, project "NarBaltAware" implemented under the European Neighbourhood Instrument and co-financed by the European Union, and Erasmus Mundus master's programme "Excellence in Analytical Chemistry (EACH)" (www.analyticalchemistry.eu, accessed on 20 November 2023) coordinated by the University of Tartu and Estonian Center of Analytical Chemistry. EcoBalt 2023 organizers also acknowledge conference supporters shown in Figure 2.

Figure 2. Supporters of EcoBalt 2023.

The session "Data Quality in Environmental Analysis" organized by the EACH program was part of the EACH mini-conference series "Data Quality in Analytical Chemistry" and additionally served as a "get-together event" for current EACH students and EACH alumni.

Acknowledgments: The publication of this volume of EcoBalt 2023 abstracts and organization of EcoBalt 2023 were financially supported by the Estonian Society of Toxicology, the Erasmus+ Pro-

gramme of the European Union (EACH EM JMD, project no. 619658), the European Neighbourhood Instrument co-financed by the European Union, as well as the supporters listed in Figure 2.

Conflicts of Interest: The authors declare no conflict of interest.

proceedings

Abstract

The Challenge of Understanding—From Protozoa to Ecosystems †

Priit Zingel

Center for Limnology, Institute of Agricultural and Environmental Sciences, Estonian University of Life Sciences, 51006 Tartu, Estonia; priit.zingel@emu.ee

† Presented at the International Conference EcoBalt 2023 "Chemicals & Environment", Tallinn, Estonia, 9–11 October 2023.

Abstract: We live in the information age, but we still do not fully understand how the surrounding nature exchanges, collects, and manages information—both at the ecosystem and cellular levels. Still, it is crucial to possess a dynamic and developing comprehension of the functioning of nature in order to understand the phenomena occurring in our ever-changing world. Sometimes we encounter an attitude that there is no need to investigate anything further, as everything has already been researched. This is a very dangerous attitude. Without information, we should not actually talk about the age of information, and it is worth reviewing some knowledge that is deeply ingrained in us from time to time. Occasionally, it is worth taking a step back and considering whether we ourselves may have become trapped in dogmas that may hinder our understanding. This presentation focuses on various aspects of understanding nature, starting from single-celled organisms and ending with ecosystems. The topics that will be discussed among others are (1) infochemicals (how do they actually affect the biota?); (2) the remarkable process of cellular computation (does this imply that cellular materials can show a primitive intelligence?); and (3) the ability of plants to influence wind patterns and bring in more moisture from the ocean (what will happen to us when we excessively deforest?).

Keywords: cellular computation; infochemicals; wind patterns

Citation: Zingel, P. The Challenge of Understanding—From Protozoa to Ecosystems. *Proceedings* **2023**, *92*, 22. https://doi.org/10.3390/proceedings2023092022

Academic Editors: Monika Mortimer, Anne Kahru, Ivo Leito, Riin Rebane and Villem Aruoja

Published: 21 November 2023

Funding: This research received no external funding.

Institutional Review Board Statement: Not applicable.

Informed Consent Statement: Not applicable.

Data Availability Statement: Not applicable.

Conflicts of Interest: The author declares no conflict of interest.

proceedings

Abstract

Sampling, Detection and Uncertainty in Environmental Analysis—Challenges and Solutions †

Stephen L. R. Ellison

LGC Limited, Queens Road, Teddington TW11 0LY, UK; s.ellison@lgcgroup.com

† Presented at the International Conference EcoBalt 2023 "Chemicals & Environment", Tallinn, Estonia, 9–11 October 2023.

Keywords: sampling uncertainty; measurement uncertainty; validation; detection capability

Analytical measurements are increasingly vital to inform our understanding of our changing global environment and to support the regulation of human activity that affects the environment. Environmental analysis, however, experiences particular challenges for measurement reliability. Most practical environmental measurements are carried out on relatively small samples and their results are taken as indicators of the much larger area sampled. Environmental monitoring can span decades, and the comparability and consistency of measurement results over time and across geographical regions is important for detecting real trends. Environmentally important contaminants—often the subject of regulations—are frequently present at very low levels, often stretching the detection capability of even today's analytical methods and instrumentation. At low levels, with significant sampling variability, and perhaps especially when environmental measurement is contentious, it is important to understand and express uncertainties clearly and accurately so that reliable policy and regulatory decisions can be made. This, with any accompanying conformity assessment decisions, can be particularly challenging in the frequent cases where sampling and even measurement distributions are far from the familiar normal distribution. Finally, the regulatory framework controlling laboratory operations has evolved over time; for example, the validation of test methods has become increasingly important as new regulatory flexibility allows wider choice of measurement methods, including 'in-house' methods, subject to achieving specific performance criteria. Here, these issues will be discussed in the light of experience of some of the UK's frameworks for environmental regulation and analysis, and with attention to some important Eurachem guidance for analytical practice.

Sampling is often the first step in many environmental measurements. It is also among the most variable, simply because environmental systems themselves are variable on scales from kilometres to centimetres. Understanding the variation due to sampling is key to developing sound sampling strategies and for planning an appropriate allocation between sampling effort and analytical work. A recently updated Eurachem guide [1] provides guidance on the determination of sampling uncertainty, including relatively simple and economical methods for assessing sampling variation using a simple method based on limited duplicate sampling and analysis [2]. The latest edition includes even more economical approaches, which use a 'staggered nested' experimental design to reduce analytical effort [3], and also includes recent methods for summarizing uncertainty when the measurement or sampling distribution is asymmetric [4].

Once sampled, attention turns to the application of an appropriate, validated, analytical method. Validation procedures for environmental analysis in the UK are specified, for example, in the MCERTS performance standard for soil analysis [5]. This sets out a specific set of validation procedures, together with criteria for acceptable performance for a wide range of analytes. A recent example of such a validation study, for an in-house modification

Citation: Ellison, S.L.R. Sampling, Detection and Uncertainty in Environmental Analysis—Challenges and Solutions. *Proceedings* **2023**, *92*, 70. https://doi.org/10.3390/proceedings2023092070

Published: 1 December 2023

of a standard method, [6] illustrated some of the problems of achieving reliable results across different soil matrices, even with very precise methods (Figure 1); clearly, some matrices can provide individual challenges (LGC6145 for nickel in Figure 1). These pose practical difficulties for achieving performance and for reporting results and uncertainty.

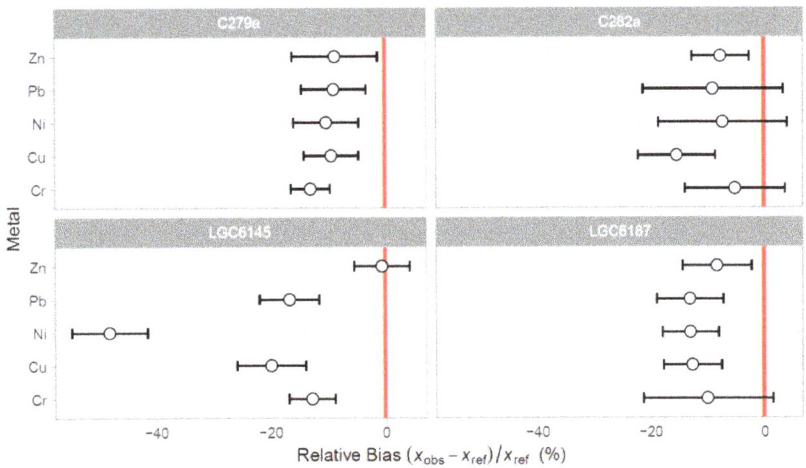

Figure 1. Relative bias for aqua regia extractable metals in four prepared soil materials. The vertical red line is at zero bias.

Detectability poses further problems, particularly for estimating averages and for summary reporting. For example, a preponderance of low levels and the wide use of 'less than' statements in summary emissions data for paper mills in the UK led to practical difficulties in assessing compliance with new environmental controls [7].

Finally, environmental analysis is not complete without a reported value, with associated measurement uncertainty and, where necessary, a statement of conformity. Measurement uncertainty evaluation is a requirement for accredited laboratories, even if it need not always be reported [8]. Originally a challenge for analytical laboratories unfamiliar with the concept, several guides exist to help with uncertainty evaluation, including a general Eurachem guide [9] and a NordTest guide specific to environmental laboratories [10]. There is also comprehensive guidance on the use of uncertainty in conformity statements, for those laboratories required to assess conformity with a requirement [11,12]. Recent studies show, however, that asymmetry of the kind found in environmental sampling distributions can adversely affect producer and consumer risks in conformity assessment [13], and some recent studies of the effect of asymmetry will be described.

Funding: This research received no external funding.

Institutional Review Board Statement: Not applicable.

Informed Consent Statement: Not applicable.

Data Availability Statement: Data sharing is not applicable to this abstract.

Conflicts of Interest: The author declares no conflict of interest.

References

1. Ramsey, M.H.; Ellison, S.L.R.; Rostron, P. (Eds.) *Eurachem/EUROLAB/ CITAC/Nordtest/AMC Guide: Measurement Uncertainty Arising from Sampling: A Guide to Methods and Approaches*, 2nd ed; Eurachem, 2019; Available online: http://www.eurachem.org (accessed on 5 October 2023).
2. Ramsey, M.H. Sampling as a source of measurement uncertainty: Techniques for quantification and comparison with analytical sources. *J. Anal. At. Spectrom.* **1998**, *13*, 97–104. [CrossRef]

3. Rostron, P.; Ramsey, M.H. Cost effective, robust estimation of measurement uncertainty from sampling using unbalanced ANOVA. *Accredit. Qual. Assur.* **2012**, *17*, 7–14. [CrossRef]
4. Ramsey, M.H.; Ellison, S.L.R. Combined uncertainty factor for sampling and analysis. *Accredit. Qual. Assur.* **2017**, *22*, 187–189. [CrossRef]
5. Environment Agency. Performance Standard for Laboratories Undertaking Chemical Testing of Soil. *Version 5*. 2018. Available online: http://www.mcerts.net/ (accessed on 5 October 2023).
6. Ellison, S.L.R.; Singh, M.; Cox, M.G. Measurement uncertainty for routine testing of metals in soil. In *Good Practice in Evaluating Measurement Uncertainty—Compendium of Examples*; van der Veen, A.M.H., Cox, M.G., Eds.; EMUE project report; EURAMET: Braunschweig, Germany, 2021. [CrossRef]
7. Ellison, S.L.; Barwick, V.J.; Williams, A. Legislative limits below detection capability. *Accredit. Qual. Assur.* **2000**, *5*, 308–313. [CrossRef]
8. *ISO 17025:2017*; General Requirements for the Competence of Testing and Calibration Laboratories. ISO: Geneva, Switzerland, 2017.
9. Ellison, S.L.R.; Williams, A. (Eds.) *Eurachem/CITAC guide: Quantifying Uncertainty in Analytical Measurement*, 3rd ed; Eurachem, 2012; Available online: www.eurachem.org (accessed on 5 October 2023).
10. Magnusson, B.; Näykki, T.; Hovind, H.; Krysell, M.; Sahlin, E. Handbook for Calculation of Measurement Uncertainty in ENVIRONMENTAL Laboratories, Nordtest Report TR 537, 4th edNordtest, 2017; Available online: www.nordtest.info (accessed on 5 October 2023).
11. *ILAC G8:2019*; Guidelines on Decision Rules and Statements of Conformity. ILAC: Silverwater, Australia, 2019.
12. Williams, A.; Magnusson, B. (Eds.) *Eurachem/CITAC Guide: Use of Uncertainty Information in Compliance Assessment*, 2nd ed; Eurachem, 2021; Available online: www.eurachem.org (accessed on 5 October 2023).
13. Kuselman, I.; Shpitzer, S.; Pennecchi, F.; Burns, C. Investigating out-of-specification test results of mass concentration of total suspended particulates in air based on metrological concepts—A case study. *Air Qual. Atmos. Health* **2012**, *5*, 269–276. [CrossRef]

proceedings

MDPI

Abstract

Nanoplastic–Biomolecular Interactions †

Pu Chun Ke [1,2]

1 Drug Delivery, Disposition and Dynamics, Monash Institute of Pharmaceutical Sciences, Monash University, 381 Royal Parade, Parkville, VIC 3052, Australia; pu-chun.ke@monash.edu
2 The Nanomedicine Center, The Great Bay Area National Institute for Nanotechnology Innovation, 136 Kaiyuan Avenue, Guangzhou 510700, China

† Presented at the International Conference EcoBalt 2023 "Chemicals & Environment", Tallinn, Estonia, 9–11 October 2023.

Keywords: nanoplastic; bio-nano interaction; endothelial leakiness; amyloidosis; environment

The global-scale production of plastics has been instrumental for sustaining the modern way of life, while the accumulation of plastics in landfills, oceans, and any other environment has become a major stressor for environmental sustainability, climate, and, potentially, human health. While mechanical and chemical forces applied by man and nature can break down and recycle plastics, our understanding of the biological fingerprints of discharged plastics, especially of the nanoscale derivatives of plastics (i.e., nanoplastics), remains superficial. In 2010, we first reported on algal photosynthesis impaired by nanoplastic adsorption [1]. More recently, a host of studies have been conducted to elucidate the environmental implications of micro- and nanoplastics at the molecular, cellular, or whole-organism level, typically from a toxicological point of view. In this paper, I will first introduce our early representative studies focused on nanoparticle–biomolecular/environmental interactions [2–6]. I will then report on our recent finding that anionic polystyrene and poly(methyl methacrylate) nanoparticles can elicit disruptions in vascular endothelial cadherin junctions, a new phenomenon that is biophysical/biochemical and uncorrelated with cytotoxic events such as reactive oxygen species production, autophagy, and apoptosis [7,8]. The last part of my presentation will be focused on the effects of nanoplastics on the aberrant aggregation of amyloid beta and alpha synuclein, two pathogenic proteins associated with Alzheimer's and Parkinson's diseases [9]. This presentation aims to demonstrate the vast research potential towards elucidating the implications of plastics for environmental sustainability and human health protection.

Citation: Ke, P.C. Nanoplastic–Biomolecular Interactions. *Proceedings* **2023**, *92*, 65. https://doi.org/10.3390/proceedings2023092065

Academic Editors: Monika Mortimer, Anne Kahru, Ivo Leito, Riin Rebane and Villem Aruoja

Published: 29 November 2023

Funding: This work was supported by the National Natural Science Foundation of China, grant number T2250710182, and by the National Key Research and Development Program, Ministry of Science and Technology of China, grant number 2022YFC2409700.

Institutional Review Board Statement: Not applicable.

Informed Consent Statement: Not applicable.

Data Availability Statement: Not applicable.

Conflicts of Interest: The author declares no conflict of interest.

References

1. Bhattacharya, P.; Lin, S.; Turner, J.P.; Ke, P.C. Physical adsorption of charged plastic nanoparticles affects algal photosynthesis. *J. Phys. Chem. C* **2010**, *114*, 16556–16561. [CrossRef]
2. Lin, S.; Reppert, J.; Hu, Q.; Hudson, J.S.; Reid, M.L.; Ratnikova, T.A.; Rao, A.M.; Luo, H.; Ke, P.C. Uptake, translocation, and transmission of carbon nanomaterials in rice plants. *Small* **2009**, *5*, 1128–1132. [CrossRef] [PubMed]
3. Qiao, R.; Roberts, A.P.; Mount, A.S.; Klaine, S.J.; Ke, P.C. Translocation of C60 and its derivaties across a lipid bilayer. *Nano Lett.* **2007**, *7*, 614–619. [CrossRef] [PubMed]
4. Ke, P.C.; Lamm, M.H. A Biophysical Perspective of Understanding Nanoparticles at Large. *Phys. Chem. Chem. Phys.* **2011**, *13*, 7273–7283. [CrossRef] [PubMed]
5. Radic, S.; Davis, T.P.; Ke, P.C.; Ding, F. Contrasting effects of nanoparticle–protein attraction on amyloid aggregation. *RSC Adv.* **2015**, *5*, 105489–105498. [CrossRef] [PubMed]
6. Li, R.; Chen, R.; Chen, P.; Wen, Y.; Ke, P.C.; Cho, S.S. Computational and Experimental Characterizations of Silver Nanoparticle–Apolipoprotein Biocorona. *J. Phys. Chem. B* **2013**, *117*, 13451–13456. [CrossRef] [PubMed]
7. Lee, M.; Ni, N.; Tang, H.; Li, Y.; Wei, W.; Kakinen, A.; Wan, X.; Davis, T.P.; Song, Y.; Leong, D.T.; et al. A Framework of Paracellular Transport via Nanoparticles-Induced Endothelial Leakiness. *Adv. Sci.* **2021**, *8*, 2102519. [CrossRef] [PubMed]
8. Wei, W.; Li, Y.; Lee, M.; Andrikopoulos, N.; Lin, S.; Chen, C.; Leong, D.T.; Ding, F.; Song, Y.; Ke, P.C. Anionic nanoplastic exposure induces endothelial leakiness. *Nat. Comm.* **2022**, *13*, 4757. [CrossRef] [PubMed]
9. Liang, X.; Andrikopoulos, N.; Tang, H.; Wang, Y.; Ding, F.; Ke, P.C. Nanoplastic Stimulates the Amyloidogenesis of Parkinson's Alpha-Synuclein NACore. *Small* **2023**, 2308753. [CrossRef] [PubMed]

proceedings

Abstract

Measurement Quality in Analysis—Guidelines and Software Tools †

Bertil Magnusson

Trollboken AB, s-41654 Gothenburg, Sweden; bertil.magnusson@trollboken.se

† Presented at the International Conference EcoBalt 2023 "Chemicals & Environment", Tallinn, Estonia, 9–11 October 2023.

Abstract: Measurement quality is about fulfilling analytical requirements, which should be based on the intended use of the results. Within the European Union, environmental requirements can be set in a directive based on the maximum allowable concentration of a substance in air, soil or water, e.g., an Environmental Quality Standard (EQS) or an Emission Limit Value (EVS); and the requirements on measurement quality, e.g., the limit of quantification (LOQ), within-laboratory reproducibility (s_{Rw}) or measurement uncertainty (MU). This presentation is about publicly available guidelines that can help the analytical chemist working in the laboratory by implementing a specific method to (1) set up internal quality control over the whole concentration range based on the requirements through the use of target control limits, (2) perform ongoing internal quality control based on only two rules, (3) plan method validation, and (4) estimate the MU based on quality control and validation data. Normally for instrumental methods, the MU is estimated as a relative uncertainty at higher concentrations and as an absolute uncertainty at lower concentrations close to the LOQ. For quality control and MU estimation, free open source software will be presented.

Keywords: measurement quality; quality control; validation; uncertainty; target control limits

Citation: Magnusson, B. Measurement Quality in Analysis—Guidelines and Software Tools. *Proceedings* **2023**, *92*, 74. https://doi.org/10.3390/proceedings2023092074

Academic Editors: Monika Mortimer, Anne Kahru, Ivo Leito, Riin Rebane and Villem Aruoja

Published: 5 December 2023

Funding: This research received no external funding.

Institutional Review Board Statement: Not applicable.

Informed Consent Statement: Not applicable.

Data Availability Statement: Data sharing not applicable.

Conflicts of Interest: The authors declare no conflict of interest.

proceedings

MDPI

Abstract

The Impact of Microplastics on Soil Invertebrates †

Anita Jemec Kokalj

Department of Biology, Biotechnical Faculty, University of Ljubljana, Jamnikarjeva 101, 1000 Ljubljana, Slovenia; anita.jemec@bf.uni-lj.si

† Presented at the International Conference EcoBalt 2023 "Chemicals & Environment", Tallinn, Estonia, 9–11 October 2023.

Keywords: microplastics; agricultural plastics; plastics-associated chemicals; terrestrial invertebrates; crustaceans; insects; soil; reproduction; growth; immune response

As a result of plastic pollution and intentional use of plastics in agriculture, small plastic particles called microplastics (<1 mm) are commonly found in soils. This has raised many concerns on the possible effects of microplastics on soil physico-chemical properties, soil biota and the ecosystem services they provide [1,2]. Hazard assessment of microplastics on soil biota relies on using model terrestrial invertebrates, which are exposed to microplastics either by ingesting the soils or through body surface contact. In recent years, we have gathered considerable information regarding the effects of microplastics on key ecotoxicity test invertebrates: woodlice *Porcellio scaber*, mealworms *Tenebrio molitor*, springtails *Folsomia candida*, enchytraeids *Enchytraeus cripticus* and earthworms *Eisenia andrei* [3–6]. The organisms were exposed through soil or food spiked with environmentally relevant microplastic concentrations (0.005–5%) for typically 3–4 weeks. Microplastics were generated from various plastic materials, generating tire wear particles, textile fibres, polypropylene microplastics from disposable medical masks, low-density polyethylene fragments from packaging, low-density-polyethylene fragments milled from mulching films and starch blend polybutylene adipate-co-terephthalate mulching film fragments. A number of endpoints was followed: survival, reproduction, moult, growth, energy-related biomarkers and immune response. In general, our results show that tested microplastics are not lethal to tested invertebrates, but microplastic exposure can induce sublethal effects, such as alternation in reproduction, reduced growth, changes in metabolic activity and induction of immune response. We will present an overview of the various effects observed on different organisms and discuss how the responses differ between types of microplastics and whether plastics-associated chemicals might contribute to the observed effects.

Citation: Kokalj, A.J. The Impact of Microplastics on Soil Invertebrates. *Proceedings* **2023**, *92*, 82. https://doi.org/10.3390/proceedings2023092082

Academic Editors: Monika Mortimer, Anne Kahru, Ivo Leito, Riin Rebane and Villem Aruoja

Published: 24 January 2024

Funding: This research was funded by: the international project IMPASSE –Impacts of MicroPlastics on AgrosyStems and Stream Environments, which is financed under the ERA-NET waterworks2015 co-funded call, Slovenian Research Agency (J1-2482, P1-0184), and European Union's Horizon 2020 project PAPILLONS (Plastic in Agricultural Production: Impacts, Life-cycle and LONg-term Sustainability) (grant agreement No 101000210).

Institutional Review Board Statement: Not applicable.

Informed Consent Statement: Not applicable.

Data Availability Statement: No new data were created or analyzed in this study. Data sharing is not applicable to this article.

Acknowledgments: I would like to acknowledge all co-authors of microplastics-related publications published in our research group. The conclusions of these publications will be presented.

Conflicts of Interest: The author declares no conflicts of interest.

References

1. de Souza Machado, A.A.; Kloas, W.; Zarfl, C.; Hempel, S.; Rillig, M.C. Microplastics as an emerging threat to terrestrial ecosystems. *Glob. Chang. Biol.* **2017**, *24*, 1405–1416. [CrossRef] [PubMed]
2. Nizzetto, L.; Futter, M.; Langaas, S. Are agricultural soils dumps for microplastics of urban origin? *Environ. Sci. Technol.* **2016**, *50*, 10777–10779. [CrossRef] [PubMed]
3. Selonen, S.; Andraz, D.; Jemec Kokalj, A.; Skalar, T.; Parramon Dolcet, L.; Hurley, R.; van Gestel, C.A.M. Exploring the impacts of plastics in soil—The effects of polyester textile fibers on soil invertebrates. *Sci. Total Environ.* **2020**, *700*, 134451. [CrossRef] [PubMed]
4. Selonen, S.; Dolar, A.; Kokalj, A.J.; Sackey, L.N.; Skalar, T.; Fernandes, V.C.; Rede, D.; Delerue-Matos, C.; Hurley, R.; Nizzetto, L.; et al. Exploring the impacts of microplastics and associated chemicals in the terrestrial environment–Exposure of soil invertebrates to tire particles. *Environ. Res.* **2021**, *201*, 111495. [CrossRef] [PubMed]
5. Kokalj, A.J.; Dolar, A.; Drobne, D.; Škrlep, L.; Škapin, A.S.; Marolt, G.; Nagode, A.; van Gestel, C.A. A Effects of microplastics from disposable medical masks on terrestrial invertebrates. *Hazmat* **2022**, *438*, 129440.
6. Jemec Kokalj, A.; Dolar, A.; Titova, J.; Visnapuu, M.; Škrlep, L.; Drobne, D.; Vija, H.; Kisand, V.; Heinlaan, M. Long term exposure to virgin and recycled LDPE microplastics induced minor effects in the freshwater and terrestrial Crustaceans *Daphnia magna* and *Porcellio scaber*. *Polymers* **2021**, *13*, 771. [CrossRef] [PubMed]

proceedings

MDPI

Abstract

Endocrine Disrupting Activity of Mixtures Composed of Pharmaceuticals and Nanoplastics †

Lucija Božičević, Nikolina Peranić, Nikolina Kalčec and Ivana Vinković Vrček *

Institute for Medical Research and Occupational Health, 10 000 Zagreb, Croatia; lbozicevic@imi.hr (L.B.); nperanic@imi.hr (N.P.); nkalcec@imi.hr (N.K.)

* Correspondence: ivinkovic@imi.hr

† Presented at the International Conference EcoBalt 2023 "Chemicals & Environment", Tallinn, Estonia, 9–11 October 2023.

Keywords: estrogen receptors; polystyrene nanoparticles; paracetamol; ibuprofen; fluoxetine; mixture effects

Endocrine disrupting chemicals (EDCs) are defined as "an exogenous chemical or mixture of chemicals that can interfere with any aspect of the hormone action" [1,2]. Thus, they have been considered among the strong risk factors in the development of obesity, metabolic disorders, infertility, endocrinopathies, diabetes, and hormone-dependent cancers globally. Combined exposure to EDCs may have more pronounced adverse effects on human health and may trigger stronger (or occasionally weaker) toxicological effects than exposure to an individual chemical, even at concentrations that are regarded as non-adverse (i.e., where no effects are expected) [3]. This study is focused on the endocrine-disruptive (ED) effects of the mixture of pharmaceuticals with nanoplastics. Paracetamol, ibuprofen, and fluoxetine were selected as scientific cases of pharmaceuticals, while commercially available 25 nm sized polystyrene nanoparticles (PNPs) were used as nanoplastics. The ED effects of each pharmaceutical and PNP, as well as of their mixtures, were evaluated using an in vitro estrogen receptor activity assay based on a T47D-KBluc cell line [4]. This cell line is stably transfected with a triplet ERE (estrogen-responsive element)-promoter-luciferase reporter gene construct, and therefore, it can be used to screen chemicals for estrogenic and anti-estrogenic effects. The obtained results showed estrogenic effects of the PNPs and all tested pharmaceuticals. The mixture of pharmaceuticals with PNPs demonstrated a higher agonistic affinity towards estrogen receptors (ERs) compared with individual components of the mixture. This study unambiguously shows that the health hazard potential of environmental contaminants should not be investigated exclusively as individual pollutants, but as complex mixture components.

Citation: Božičević, L.; Peranić, N.; Kalčec, N.; Vrček, I.V. Endocrine Disrupting Activity of Mixtures Composed of Pharmaceuticals and Nanoplastics. *Proceedings* **2023**, *92*, 1. https://doi.org/10.3390/proceedings2023092001

Academic Editor: Monika Mortimer, Anne Kahru, Ivo Leito, Riin Rebane and Villem Aruoja

Published: 20 November 2023

Author Contributions: Conceptualization, I.V.V.; methodology, L.B., N.P. and N.K.; validation, L.B.; formal analysis, L.B., N.P. and N.K.; investigation, L.B.; resources, I.V.V.; data curation, L.B., N.P. and N.K.; writing—original draft preparation, L.B., N.P. and N.K.; writing—review and editing, I.V.V.; visualization, L.B., N.P. and N.K.; supervision, I.V.V.; project administration, I.V.V.; funding acquisition, I.V.V. All authors have read and agreed to the published version of the manuscript.

Funding: This research was funded by the Croatian-Chinese bilateral project "Endocrine disrupting mechanism of typical environmental pollutants—EmergeTox" funded by the Ministry of Science and Education, Republic of Croatia and by the Chinese Academy of Sciences and the CRO-SAD 2/2019 bilateral project and by the EU H2020 project (H2020-NMBP-13-2018 RIA): RiskGONE (Science-based Risk Gover-nance of NanoTechnology) under grant agreement n814425.

Institutional Review Board Statement: Not applicable.

Informed Consent Statement: Not applicable.

Data Availability Statement: Data sharing not applicable.

Conflicts of Interest: The authors declare no conflict of interest.

References

1. Schug, T.T.; Johnson, A.F.; Birnbaum, L.S.; Colborn, T.; Guillette Jr, L.J.; Crews, D.P.; Collins, T.; Soto, A.M.; Vom Saal, F.S.; McLachlan, J.A.; et al. Minireview: Endocrine Disruptors: Past Lessons and Future Directions. *Mol. Endocrinol.* **2016**, *30*, 833–847. [CrossRef] [PubMed]
2. Lauretta, R.; Sansone, A.; Sansone, M.; Romanelli, F.; Appetecchia, M. Endocrine Disrupting Chemicals: Effects on Endocrine Glands. *Front. Endocrinol.* **2019**, *10*, 178. [CrossRef] [PubMed]
3. Available online: https://publications.jrc.ec.europa.eu/repository/bitstream/JRC123531/jrc123531online.pdf (accessed on 30 May 2023).
4. USEPA, Office of Prevention P and TS: Endocrine Disruptor Screening Program Test Guidelines—OPPTS 890.1300: Estrogen Receptor Transcriptional Activation. 2009. Available online: https://nepis.epa.gov/Exe/ZyPURL.cgi?Dockey=P100SHG1.txt (accessed on 30 May 2023).

proceedings

MDPI

Abstract

Chasing Pollutants Concerning Public Health: From Food to Smoke †

Urška Šunta, Polonca Trebše , Borut Poljšak and Mojca Bavcon Kralj *

Faculty of Health Sciences, University of Ljubljana, 1000 Ljubljana, Slovenia; urska.sunta@zf.uni-lj.si (U.Š.); polonca.trebse@zf.uni-lj.si (P.T.); borut.poljsak@zf.uni-lj.si (B.P.)
* Correspondence: mojca.kralj@zf.uni-lj.si; Tel.: +386-1-300-11-11

† Presented at the International Conference EcoBalt 2023 "Chemicals & Environment", Tallinn, Estonia, 9–11 October 2023.

Keywords: solid-phase microextraction; food aromas; microplastics identification; smoking comparison; volatile organic compounds

Sample preparation is the key step in determining low concentrations of pollutants from food, biological, plant, industrial, and environmental matrices. Solid-phase microextraction (SPME) is a solvent-free, cost-effective, robust, and high-throughput sample preparation technique [1] usually coupled with gas chromatography (GC). It is especially versatile for sampling volatile organic compounds (VOCs) not only present in the analysis of wine bouquet or cheese aromas [2], but also in cigarette smoke [3], or in the identification of microplastics (MPs) [4]. The sorption of VOCs on the SPME fibre in the headspace (HS) of the sample depends on numerous parameters: the type of fibre, the extraction time, and the temperature. Mixed-polarity phase SPME fibres (DVB/CAR/PDMS; Supelco, Bellefonte, PA, USA) were used in all analyses, from Nanos cheese to MP identification to cigarette smoke. The HS-SPME method enabled the VOCs' profile study of Nanos cheese. The evolved cheese aroma profiles were affected by cheesemaking parameters: the amount of starter culture, ripening temperature, and media, and were independent of the geographical origin of raw milk as well as the location of ripening [2]. Further, by employing the HS-SPME-GC–MS method, identification of the five most common polymer types (PVC, PS, PET, PP, and PE) of MPs was possible. The well-controlled melting process, which generates characteristic compounds of each polymer, enabled the classification of MPs from real mixtures. Studying other VOCs concerning public health also included HS-SPME of flavours in tobacco products [classic cigarettes (CCs), electronic cigarettes (ECs), and heat-no-burn products (HNB)]. Flavours are the most common reason for promoting smoking initiation and duration, and they make smoking cessation more difficult among adolescents. However, the lack of simple smoke/aerosol/vapour (S/A/V) analyses for comparison of CCs, ECs, and HNB makes legislation or prohibition of such products impossible. It would be of general interest if a simple, standardised method existed.

Citation: Šunta, U.; Trebše, P.; Poljšak, B.; Kralj, M.B. Chasing Pollutants Concerning Public Health: From Food to Smoke. *Proceedings* **2023**, *92*, 60. https://doi.org/10.3390/proceedings2023092060

Academic Editors: Monika Mortimer, Anne Kahru, Ivo Leito, Riin Rebane and Villem Aruoja

Published: 29 November 2023

Author Contributions: Conceptualisation, M.B.K.; methodology, U.Š. and M.B.K.; validation, U.Š.; formal analysis, M.B.K. and U.Š.; investigation, M.B.K., P.T., U.Š. and B.P.; resources, P.T.; data curation, M.B.K.; writing—original draft preparation, M.B.K., P.T., U.Š. and B.P.; writing—review and editing, M.B.K., P.T. and U.Š.; visualisation, M.B.K.; supervision, M.B.K. and P.T.; project administration, P.T. and B.P.; funding acquisition, M.B.K., P.T. and B.P. All authors have read and agreed to the published version of the manuscript.

Funding: This research was funded by ARRS P3-0388 for support through the programme "Mechanisms of Health Maintenance" and the post-graduate research funding programme Young Researchers (SP-0510/21 and 100-22-0510).

Institutional Review Board Statement: Not applicable.

Informed Consent Statement: Not applicable.

Data Availability Statement: Not applicable.

Acknowledgments: The authors acknowledge Iva Boltar for contributing to this research work.

Conflicts of Interest: The authors declare no conflict of interest.

References

1. Jalili, V.; Barkhordari, A.; Ghiasvand, A. A comprehensive look at solid-phase microextraction technique: A review of reviews. *Microchem. J.* **2020**, *152*, 104319. [CrossRef]
2. Boltar, I.; Majhenič, A.Č.; Jarni, K.; Jug, T.; Kralj, M.B. Research of volatile compounds in cheese affected by different technological parameters. *J. Food Nutr. Res.* **2019**, *58*, 75–84.
3. Poljšak, B.; Gošnak Dahmane, R.; Podobnik, L.; Bavcon Kralj, M. Comparing smoke and aerosols from classic and electronic cigarettes, and from «heat-not-burn» devices. In *Proceedings of the 22nd European Meeting on Environmental Chemistry: Book of Abstracts, Ljubljana, Slovenia, 5–8 December 2022*, 1st ed.; Bavcon Kralj, M., Trebše, P., Prosenc, F., Šunta, U., Čižmek, L., Eds.; University of Ljubljana Press: Ljubljana, Slovenia, 2022; ISBN 978-961-297-035-2.
4. Šunta, U.; Trebše, P.; Kralj, M.B. Simply applicable method for microplastics determination in environmental samples. *Molecules* **2021**, *26*, 1840. [CrossRef] [PubMed]

proceedings

Abstract

Role of Plant Volatiles in Atmospheric Processes under Current and Future Climates †

Ülo Niinemets

Department of Crop Science and Plant Biology, Estonian University of Life Sciences, Kreutzwaldi 1, 51011 Tartu, Estonia; ylo.niinemets@emu.ee; Tel.: +372-53457189

† Presented at the International Conference EcoBalt 2023 "Chemicals & Environment", Tallinn, Estonia, 9–11 October 2023.

Abstract: Several plant species emit different volatile organic compounds constitutively under typical physiological conditions, and all plant species can be elicited to emit volatiles under stress conditions when physiological activity is curbed. Plant emissions constitute the biggest global source of reactive volatile organic compounds in the atmosphere. These volatiles play major biological roles in plant–plant interactions and in plant interactions with other organisms. In addition, plant-emitted volatiles often dominate the volatile-driven large-scale biosphere–atmosphere processes. In particular, plant volatiles participate in ozone- and secondary organic aerosol (SOA)-forming reactions and in cloud formation. Both enhanced SOA and cloud formation contribute to the cooling of the Earth's surface, implying that plants can alter their own life environment. At the global scale, amelioration of vegetation growing conditions via volatile emissions can reduce the rate of global environmental change, but there are currently major uncertainties in understanding how plant emissions change in future climates and the quantitative impact of plant emissions on future climate change. Climate change is a complex phenomenon that entails alterations in a series of environmental factors. For plants, climate change is expected to enhance the severity and duration of stress periods when plant physiological activity is strongly reduced. This includes both the enhancement of abiotic stresses such as rising temperatures and more severe drought episodes in many parts of the globe and more devastating biotic stresses such as frequent outbreaks of herbivore and pathogen attacks. These changes in the frequency and duration of stress episodes can strongly impact volatile emissions. Stress typically reduces the emissions of constitutively released plant volatiles and elicits emissions of specific volatiles in a stress-severity-dependent manner. Thus, the plant-dependent feedback on global change is expected to become stronger in more stress-prone climates.

Keywords: air chemistry; biotic stress; cloud formation; environmental stress; global change; plant stress; plant volatiles; secondary organic aerosols; stress-induced volatiles; volatile organic compounds

Citation: Niinemets, Ü. Role of Plant Volatiles in Atmospheric Processes under Current and Future Climates. *Proceedings* **2023**, *92*, 73. https://doi.org/10.3390/proceedings2023092073

Academic Editors: Monika Mortimer, Anne Kahru, Ivo Leito, Riin Rebane and Villem Aruoja

Published: 5 December 2023

Funding: This research was funded by the European Research Council project SIP-VOL+, Grant No. 322603.

Institutional Review Board Statement: Not applicable.

Informed Consent Statement: Not applicable.

Data Availability Statement: The data underlying this presentation are available from the author upon request.

Conflicts of Interest: The author declares no conflict of interest.

proceedings

Abstract

The Use of Biomarkers in Monitoring and Assessment of Chemical Contamination in the Baltic Sea †

Kari K. Lehtonen

Marine and Freshwater Solutions Unit, Finnish Environment Institute (Syke), FI-00790 Helsinki, Finland; kari.lehtonen@syke.fi; Tel.: +358-(0)295-251-359

† Presented at the International Conference EcoBalt 2023 "Chemicals & Environment", Tallinn, Estonia, 9–11 October 2023.

Abstract: Chemical pollution affects the health of marine biota, key functions of the ecosystem, and endangers its biodiversity. The current environmental monitoring and assessment approach is based on chemical and ecological measurements, while the linkage between them, biological effects at the individual level, has been largely neglected. It is currently well acknowledged that the monitoring of chemical concentrations alone is not sufficient to protect populations; it considers only a tiny number of substances, while myriad others are left unnoticed. In addition, the hazards related to contaminant mixtures remain undetected. Biological effects of contaminants can be detected and measured at different levels of biological organization, i.e., from the molecular/biochemical level up to changes in populations and communities. In Baltic Sea monitoring programs, the few observations on the effects of contaminants have, for decades, mostly been made at only the higher biological organization levels, applying reproduction success parameters in top predators such as raptor birds and seals as indicators. However, recording early warning signals of contaminant exposure and effects on the health of individuals at lower biological levels makes it possible to anticipate and prevent damage at the higher levels. Moreover, it is also crucial to monitor effects at lower levels of the marine food web. The so-called biomarkers focus on changes in various biological functions and include parameters related to the detoxification of xenobiotics, oxidative stress, neurotoxicity, geno- and cytotoxicity and reproductive disorders, among others. Research on biomarkers aiming at improving their use in monitoring and assessing the contamination status of marine areas has been intensive during recent decades, although the implementation of the methods to monitoring programmes has been slow. Here, the current state of biological effects monitoring using biomarkers, as well as selected case studies in the Baltic Sea, are presented, followed by recommendations on their application in this sea area.

Keywords: Baltic Sea; biological effects; biomarkers; monitoring; chemical contaminants

Citation: Lehtonen, K.K. The Use of Biomarkers in Monitoring and Assessment of Chemical Contamination in the Baltic Sea. *Proceedings* **2023**, *92*, 80. https://doi.org/10.3390/proceedings2023092080

Academic Editors: Monika Mortimer, Anne Kahru, Ivo Leito, Riin Rebane and Villem Aruoja

Published: 19 January 2024

Funding: This research received no external funding.

Institutional Review Board Statement: Not applicable.

Informed Consent Statement: Not applicable.

Data Availability Statement: Not applicable.

Conflicts of Interest: The author declares no conflict of interest.

proceedings

MDPI

Abstract

A Chimie Douce Route to Layered Double Hydroxides †

Denis Sokol and Aivaras Kareiva *

Institute of Chemistry, Vilnius University, Naugarduko 24, LT-03225 Vilnius, Lithuania; denis.sokol@chf.vu.lt
* Correspondence: aivaras.kareiva@chgf.vu.lt; Tel.: +370-61567428

† Presented at the International Conference EcoBalt 2023 "Chemicals & Environment", Tallinn, Estonia, 9–11 October 2023.

Keywords: layered double hydroxides; sol-gel synthesis; optical properties

Recently, layered double hydroxides (LDHs) have attracted substantial attention due to their wide range of important application areas, e.g., catalysis, photochemistry, biomedical science and the environment [1,2]. LDHs can be fabricated through different synthesis methods. The most common preparation techniques are co-precipitation [3] and anion exchange [4]. The aim of this study is to show the advantages of the Chimie Douce route to LDHs. The indirect sol-gel synthesis route for the preparation of LDHs was recently developed and suggested [5]. Synthesized precursor gels were converted to mixed metal oxides (MMOs) by heating the gels at 650 °C. The LDHs were fabricated by reconstruction of MMOs in water at 80 °C. In this study, the co-precipitation and novel indirect sol-gel synthesis techniques for the preparation of Mg-Al LDHs were compared and luminescent properties have been investigated. The peculiarities of the intercalation of organic anions to the LDH structures were also studied. In conclusion, the proposed sol-gel synthesis route for LDHs shows some benefits over the co-precipitation method such as simplicity, high homogeneity and good crystallinity of the end synthesis products, effectiveness, cost efficiency and suitability for different systems. It was also demonstrated that the luminescence of lanthanide element in the $Mg_3Al_{1-x}RE_x$ could be induced by intercalation of organic reagents to the LDH structure. The Mg_3Al LDH coatings have also been successfully fabricated using the same sol-gel processing route.

Citation: Sokol, D.; Kareiva, A. A Chimie Douce Route to Layered Double Hydroxides. *Proceedings* **2023**, 92, 46. https://doi.org/10.3390/proceedings2023092046

Academic Editors: Monika Mortimer, Anne Kahru, Ivo Leito, Riin Rebane and Villem Aruoja

Published: 24 November 2023

Author Contributions: Conceptualization, A.K.; methodology, A.K. and D.S.; formal analysis, D.S.; investigation, D.S.; writing—original draft preparation, A.K.; writing—review and editing, A.K. All authors have read and agreed to the published version of the manuscript.

Funding: This research received no external funding.

Institutional Review Board Statement: Not applicable.

Informed Consent Statement: Informed consent was obtained from all subjects involved in the study.

Data Availability Statement: Not applicable.

Acknowledgments: We would like to thank A. Smalenskaite for assistance and discussions.

Conflicts of Interest: The authors declare no conflict of interest.

References

1. Mohanty, U.A.; Sahoo, D.P.; Paramanik, L.; Parida, K. A critical review on layered double hydroxide (LDH)-derived functional nanomaterials as potential and sustainable photocatalysts. *Sustain. Energy Fuels* **2023**, *7*, 1145–1186. [CrossRef]
2. Constantino, V.R.L.; Figueiredo, M.P.; Magri, V.R.; Eulalio, D.; Cunha, V.R.R.; Alcantara, A.C.S.; Perotti, G.F. Biomaterials Based on Organic Polymers and Layered Double Hydroxides Nanocomposites: Drug Delivery and Tissue Engineering. *Pharmaceutics* **2023**, *15*, 413. [CrossRef] [PubMed]
3. Theiss, F.L.; Ayoko, G.A.; Frost, R.L. Synthesis of layered double hydroxides containing Mg^{2+}, Zn^{2+}, Ca^{2+} and Al^{3+} layer cations by co-precipitation methods-A review. *Appl. Surf. Sci.* **2016**, *383*, 200–213. [CrossRef]
4. Kameda, T.; Takaizumi, C.; Kumagai, S.; Saito, Y.; Yoshioka, T. Preparation of Zn-Al layered double hydroxide intercalated with carboxymethyl-beta-cyclodextrin by anion exchange method and its Ni^{2+} adsorption property. *Soft Mater.* **2020**, *19*, 139–147. [CrossRef]
5. Sokol, D.; Salak, A.N.; Ferreira, M.G.S.; Beganskiene, A.; Kareiva, A. Bi-substituted $Mg_3Al\text{-}CO_3$ layered double hydroxides. *J. Sol-Gel Sci. Technol.* **2018**, *85*, 221–230. [CrossRef]

proceedings

MDPI

Abstract

The Environmental Safety Aspects of Technologically Powerful Materials Are Often Overlooked †

Anne Kahru [1,2,*], Margit Heinlaan [1], Irina Blinova [1], Villem Aruoja [1], Kaja Kasemets [1], Asya Drenkova-Tuhtan [1], Mariliis Sihtmäe [1] and Monika Mortimer [1]

1 Laboratory of Environmental Toxicology, National Institute of Chemical Physics and Biophysics, 23 Akadeemia Tee, 12618 Tallinn, Estonia; margit.heinlaan@kbfi.ee (M.H.); irina.blinova@kbfi.ee (I.B.); villem.aruoja@kbfi.ee (V.A.); kaja.kasemets@kbfi.ee (K.K.); asya.drenkova-tuhtan@kbfi.ee (A.D.-T.); mariliis.sihtmae@kbfi.ee (M.S.); monika.mortimer@kbfi.ee (M.M.)

2 Estonian Academy of Sciences, 6 Kohtu, 10130 Tallinn, Estonia

* Correspondence: anne.kahru@kbfi.ee

† Presented at the International Conference EcoBalt 2023 "Chemicals & Environment", Tallinn, Estonia, 9–11 October 2023.

Keywords: environmental hazard; heavy metals; rare earth elements; green technologies; green deal; sustainable-by-design; one health

Novel materials and their combinations are the basis of societal progress: stone—Stone Age; bronze—Bronze Age; iron—Iron Age. For the current stage of the development of mankind, there is not yet a commonly agreed strategic key material (silicon, polymers, graphene, nano*) but it is generally agreed that novel materials and their combinations are creating the basis of the technological progress. Not all technologically powerful materials are intrinsically safe and may harm humans and our surrounding ecosystems already at relatively low concentrations (copper, silver, zinc, cadmium, lead, mercury, nickel, chromium, platinum, lithium, and cobalt). That does not mean that intrinsically harmful materials cannot be harnessed to offer mankind new developments (incl. for the generation of green energy and the destruction of environmental pollutants). Indeed, some of the intrinsically harmful materials highly conduct electricity (copper), have magnetic properties (cobalt, nickel, and neodymium), or possess multiple technologically beneficial properties (graphene). However, progress cannot be built upon threatening the health of people and the environment. To find a balance between venture and precaution, the environmental fate and safety aspects of technologically powerful materials can no longer be overlooked to be in line with the UN Sustainable Development Goals, The Green Deal, and One Health programs. In addition, environmental toxicity data are imperatively needed for all materials sold or marketed in Europe in large quantities, as regulated by REACH legislation. Moreover, the data on the toxicity of almost all elements in the periodic table as well as on plenty of organic compounds to conduct the initial risk assessment are available in various databases and scientific resources. Unfortunately, the communities of material scientists and engineers who create novel materials and devices and environmental scientists who have knowledge on harmful effects of materials are educated separately and do not share the same information space in their professional life. Due to that, there is a big risk that the novel technologies will be introduced on a large scale before their environmental aspects (but also human health aspects) have been deeply evaluated. To avoid that, a holistic approach, covering also safety aspects [1], is needed while novel technologies are planned and designed, analogous to that applied in nanomaterials safety research about 20 years ago when physicists, biologists, chemists, material scientists, environmental scientists, and medical doctors joined forces for the analysis of potential harmful effects of nanomaterials—cornerstones for nanotechnologies [2,3].

Citation: Kahru, A.; Heinlaan, M.; Blinova, I.; Aruoja, V.; Kasemets, K.; Drenkova-Tuhtan, A.; Sihtmäe, M.; Mortimer, M. The Environmental Safety Aspects of Technologically Powerful Materials Are Often Overlooked. *Proceedings* **2023**, *92*, 11. https://doi.org/10.3390/proceedings2023092011

Academic Editors: Ivo Leito and Riin Rebane

Published: 21 November 2023

Author Contributions: Conceptualization, A.K.; writing—original draft preparation, A.K.; review and editing, M.H., I.B., V.A., K.K., M.M., M.S. and A.D.-T. All authors have read and agreed to the published version of the manuscript.

Funding: This research was funded through Estonian Research Council's grants PRG749, PRG1427, and STP28, and by the EC, Grant agreement ID: 867457, through an individual fellowship within the Marie Skłodowska-Curie MSCA-IF-EF-ST.

Institutional Review Board Statement: Not applicable.

Informed Consent Statement: Not applicable.

Data Availability Statement: Not applicable.

Conflicts of Interest: The authors declare no conflict of interest. The funders had no role in the design of the study; in the collection, analyses, or interpretation of data; in the writing of the manuscript; or in the decision to publish the results.

References

1. Bondarenko, O.; Mortimer, M.; Kahru, A.; Feliu, N.; Javed, I.; Kakinen, A.; Lin, S.; Xia, T.; Song, Y.; Davis, T.P.; et al. Nanotoxicology and nanomedicine: The Yin and Yang of nano-bio interactions for the new decade. *Nano Today* **2021**, *39*, 101184. [CrossRef] [PubMed]
2. Kahru, A.; Dubourguier, H.-C. From ecotoxicology to nanoecotoxicology. *Toxicology* **2010**, *269*, 105–119. [CrossRef] [PubMed]
3. Kahru, A.; Ivask, A. Mapping the dawn of nanoecotoxicological research. *ACC Chem. Res.* **2013**, *46*, 823–833. [CrossRef] [PubMed]

proceedings

Abstract

Wood Chemistry Perspectives at TalTech †

Tiit Lukk

Department of Chemistry and Biotechnology, Tallinn University of Technology, 19086 Tallinn, Estonia; tiit.lukk@taltech.ee; Tel.: +372-620-2005

† Presented at the International Conference EcoBalt 2023 "Chemicals & Environment", Tallinn, Estonia, 9–11 October 2023.

Abstract: Despite being the most abundant renewable resource in Estonia, wood has not been fully utilized for its potential value. To fully harness its potential, wood needs to be processed to its core components, such as fibers, wood sugars, and lignin, using chemical, thermochemical, or enzymatic methods. Researchers at TalTech are working on developing a comprehensive value chain for wood valorization, encompassing the production of food additives, thermoplastics, coating materials, and fine chemicals. Traditionally, lignin, a byproduct of the wood fractionation, was burned for energy. However, TalTech researchers are now using lignin to create novel thermoplastics, catalytic materials, and fine chemicals via fermentation processes. Here we show that the conversion of lignin into novel functional materials can be tailored by selecting appropriate extraction methods and source biomass. For instance, organosolv lignin extracted from barley straw exhibits the largest BET active surface area for lignin-based aerogels compared to those produced from softwood or hardwood organosolv lignins. Additionally, we showcase the potential of actinobacterial enzymes to detoxify phenol-containing wood sugar solutions, enabling their utilization in fermentation processes involving a wider array of microbial species.

Keywords: wood chemistry; lignin; functional materials; fermentation; environmental microbiology; TalTech

Citation: Lukk, T. Wood Chemistry Perspectives at TalTech. *Proceedings* **2023**, *92*, 86. https://doi.org/10.3390/proceedings2023092086

Academic Editors: Monika Mortimer, Anne Kahru, Ivo Leito, Riin Rebane and Villem Aruoja

Published: 29 January 2024

Funding: This study was supported by ERDF and Estonian Research Council via project RESTA11.

Institutional Review Board Statement: Not applicable.

Informed Consent Statement: Not applicable.

Data Availability Statement: No new data were created or analyzed in this abstract. Data sharing is not applicable to this abstract.

Conflicts of Interest: The author declares no conflicts of interest.

proceedings

MDPI

Abstract

What Phytoplankton Species Can Tell Us about the Implications of Engineered Nanoparticles in the Aquatic Environment †

Vera I. Slaveykova

Environmental Biogeochemistry and Ecotoxicology, Department F.-A. Forel for Environmental and Aquatic Sciences, Faculty of Sciences, University of Geneva, 66 Blvd Carl-Vogt, 1211 Geneva, Switzerland; vera.slaveykova@unige.ch

† Presented at the International Conference EcoBalt 2023 "Chemicals & Environment", Tallinn, Estonia, 9–11 October 2023.

Keywords: nanomaterials; phytoplankton; metabolomics

Nanotechnology is considered as the "sixth truly revolutionary technology" introduced into the modern world. The central question of how to benefit from this powerful technology, while maximizing and avoiding possible risks, represents a challenge for regulatory agencies and an important area of scientific research. With examples from our own research, we will illustrate what happens when aquatic microorganisms are inadvertently exposed to engineered nanoparticles (ENPs) that are increasingly released into the environment. The specific focus will be on nanoAg and nanoTiO_2, as representatives of the most widely used nanomaterials. We compared the ENP-induced responses in two phytoplankton species: the presumably "particle-proof" green alga *Chlamydomonas reinhardtii* and the "particle-ingesting" microalgal predator flagellate *Poterioochromonas malhamensis*. Generation of the reactive oxygen species (ROS), disturbing the cellular pro- and antioxidant equilibrium, as well as membrane damage and effect of ENPs on the photosynthesis were followed. The results revealed a significant increase in the cellular ROS and membrane damage upon exposure to ENPs, but the intensity of the effects was dependent on the nature, size and concentration of the ENPs, the exposure duration and the feeding pattern of the phytoplankton species. Liquid chromatography-based targeted metabolomics revealed that the abundance of metabolites involved in various pathways corresponding to amino acid, nucleotides, fatty acids, tricarboxylic acid cycle and antioxidant metabolism was altered in various treatments. The metabolomics results correlated well with the physiological results and confirmed that (i) oxidative stress is a major toxicity mechanism for nanoTiO_2 exposure [1]; and (ii) dissolved Ag released by nanoAg seems to be a major toxicity driver, even though nanoAg is internalized in the food vacuoles of *P. malhamansis* [2]. However, nanoAg plays an important role in the perturbation of amino acid metabolism, TCA cycle and oxidative stress. The implications of the obtained results for assessing the ENP-induced toxicity and tolerance responses in phytoplankton and for enabling the discovery of sensitive markers for early warning are highlighted.

Citation: Slaveykova, V.I. What Phytoplankton Species Can Tell Us about the Implications of Engineered Nanoparticles in the Aquatic Environment. *Proceedings* **2023**, *92*, 7. https://doi.org/10.3390/proceedings2023092007

Academic Editors: Monika Mortimer, Anne Kahru, Ivo Leito, Riin Rebane and Villem Aruoja

Published: 21 November 2023

Funding: This research was funded by Swiss National Science Foundation (SNSF), grant number 180186 and 204174.

Institutional Review Board Statement: Not applicable.

Informed Consent Statement: Not applicable.

Data Availability Statement: Not applicable.

Conflicts of Interest: The author declares no conflict of interest. The funder had no role in the design of the study; in the collection, analyses, or interpretation of data; in the writing of the manuscript; or in the decision to publish the results.

References

1. Liu, W.; Li, M.; Li, W.; Keller, A.A.; Slaveykova, V.I. Metabolic alterations in alga *Chlamydomonas reinhardtii* exposed to $nTiO_2$ materials. *Environ. Sci. Nano* **2022**, *9*, 2922–2938. [CrossRef] [PubMed]
2. Liu, W.; Majumdar, S.; Li, W.; Keller, A.A.; Slaveykova, V.I. Metabolomics for early detection of stress in freshwater alga *Poterioochromonas malhamensis* exposed to silver nanoparticles. *Sci. Rep.* **2020**, *10*, 20563. [CrossRef] [PubMed]

proceedings

MDPI

Abstract

The Activity of Nanomaterials in Photocatalysis †

Marina Krichevskaya

Laboratory of Environmental Technology, Department of Materials and Environmental Technology, Tallinn University of Technology, Ehitajate tee 5, 19086 Tallinn, Estonia; marina.kritsevskaja@taltech.ee

† Presented at the International Conference EcoBalt 2023 "Chemicals & Environment", Tallinn, Estonia, 9–11 October 2023.

Keywords: photocatalysis; titanium dioxide; nanoparticles; air purification; hydrogen generation

Several environmental issues need to be addressed, including air pollution and human exposure to the low concentrations of volatile organic compounds that have been detected indoors. Replacing fossil fuels with, for example, green and renewable hydrogen to reduce carbon dioxide emissions and air pollution is another recent challenge. Photocatalysis is an attractive technology with the potential to do both: abate air pollutants and synthesize hydrogen. In order to increase the potential of the environmental applications of photocatalysis, the synthesis of catalytic materials with significantly higher activity is a priority. The semiconductor oxide TiO_2 in the form of an immobilized nano-powder and in the form of thin films is the most studied and most promising photocatalyst. The photocatalytic activity of TiO_2 is associated with its defects, where both types of charge carriers—electrons and holes—can reach the surface in nanosized materials to form effective interactions [1]. Favorable defect distribution can be determined by the presence of specific facets, although obtaining high-purity anatase crystals with controlled crystallographic facet purity remains a challenge [2]. It has also been found that oxygen vacancies are critical for the adsorption of oxygen on the particle surface and the capture of photogenerated electrons. However, the engineering of oxygen vacancies has not yet been developed for practical applications of photocatalytic air purification or hydrogen generation [3,4]. At present, there is still a gap in the efficient use of photocatalytic nanomaterials between laboratory-scale research and practical applications. The development of materials science in the field of nanomaterials, along with developments in the field of photocatalytic systems engineering, should provide solutions for increasing the efficiency of photocatalysis.

Funding: This research received no external funding.

Institutional Review Board Statement: Not applicable.

Informed Consent Statement: Not applicable.

Data Availability Statement: Data sharing is not applicable to this abstract.

Conflicts of Interest: The authors declare no conflict of interest.

Citation: Krichevskaya, M. The Activity of Nanomaterials in Photocatalysis. *Proceedings* **2023**, *92*, 23. https://doi.org/10.3390/proceedings2023092023

Academic Editors: Monika Mortimer, Anne Kahru, Ivo Leito, Riin Rebane and Villem Aruoja

Published: 21 November 2023

References

1. Beni, A.A.; Jabbari, H. Nanomaterials for environmental applications. *Results Eng.* **2022**, *15*, 100467. [CrossRef]
2. Chen, F.; Cao, F.; Li, H.; Bian, Z. Exploring the important role of nanocrystals orientation in TiO_2 superstructure on photocatalytic performances. *Langmuir* **2015**, *31*, 3494–3499. [CrossRef]
3. Moiseev, A.; Krichevskaya, M.; Qi, F.; Weber, A.P.; Deubener, J. Analysis of photocatalytic performance of nanostructured pyrogenic titanium dioxide powders in view of their polydispersity and phase transition: Critical anatase particle size as a factor for suppression of charge recombination. *Chem. Eng. J.* **2013**, *228*, 614–621. [CrossRef]
4. Lu, Y.; Chen, M.; Jiang, L.; Cao, J.; Li, H.; Lee, S.C.; Huang, Y. Oxygen vacancy engineering of photocatalytic nanomaterials for enrichment, activation, and efficient removal of nitrogen oxides with high selectivity: A review. *Environ. Chem. Lett.* **2022**, *20*, 3905–3925. [CrossRef]

proceedings

MDPI

Abstract

Pulsed Corona Discharge Plasma Combined with Photocatalytic Oxidation Technology for the Degradation of Volatile Organic Compounds in Air †

Juri Bolobajev *, Kristen Altof, Marina Krichevskaya and Sergei Preis

Department of Materials and Environmental Technology, Tallinn University of Technology, Ehitajate tee 5, 19086 Tallinn, Estonia; kristen.altof@taltech.ee (K.A.); marina.kritsevskaja@taltech.ee (M.K.); sergei.preis@taltech.ee (S.P.)

* Correspondence: juri.bolobajev@taltech.ee

† Presented at the International Conference EcoBalt 2023 "Chemicals & Environment", Tallinn, Estonia, 9–11 October 2023.

Abstract: The anthropogenic impact on the environment has long been known to negatively affect the quality of air. Volatile organic compounds (VOCs) are widely used in domestic and industrial applications, generally as solvents. They are mobile in both gaseous and aqueous phases, and thus their spread in environment could have massive effect with dramatically negative consequences. Pulsed corona discharge (PCD) and photocatalytic oxidation (PCO) are considered as efficient and eco-friendly methods for the energy-efficient abatement of gaseous hazardous pollutants. One of the main problems of PCD application in air treatment, however, is residual ozone, a side product of air ionization considered as secondary air pollution. Photocatalytic processes are known to degrade ozone extending simultaneously the photocatalyst lifetime. Thus, combining PCD and PCO in a two-step treatment system could solve the problem of the presence of residual ozone and complement each other's strengths. In this study, experiments were conducted in separate systems, i.e. photocatalysis and plasma, making a prerequisite for the progress in the combined PCD/PCO applications. A prototype PCO reactor was built and tested with ozone and 2-methoxyethanol (2ME) in combinations. 2ME was chosen as a hazardous model VOC used in industry in solvents and paints. For the PCD experiments xylene was tested. Being refractory air pollutant, extensively studied for its removal, xylene provides a basis for the comparison of its abatement methods. The PCD treatment showed unequalled energy efficiencies in gaseous xylene oxidation. With respect to PCO experiments, the degradation of 2ME and ozone was 40% and 95%, respectively. High ozone degradation performed by PCO confirms the expediency of proposed air cleaning combination.

Keywords: volatile organic pollutant; heterogeneous photocatalysis; pulsed corona discharge

Citation: Bolobajev, J.; Altof, K.; Krichevskaya, M.; Preis, S. Pulsed Corona Discharge Plasma Combined with Photocatalytic Oxidation Technology for the Degradation of Volatile Organic Compounds in Air. *Proceedings* **2023**, *92*, 81. https://doi.org/10.3390/proceedings2023092081

Academic Editors: Monika Mortimer, Anne Kahru, Ivo Leito, Riin Rebane and Villem Aruoja

Published: 24 January 2024

Author Contributions: Conceptualization, J.B., K.A. and M.K.; methodology, J.B. and K.A.; formal analysis, J.B., K.A., M.K. and S.P.; investigation, J.B. and K.A.; resources, S.P.; writing—original draft preparation, J.B.; writing—review and editing, J.B., K.A. and S.P.; supervision, J.B. and M.K.; funding acquisition, S.P. All authors have read and agreed to the published version of the manuscript.

Funding: This work was supported by the Institutional Development Program of Tallinn University of Technology for 2016–2022, project 2014-2020.4.01.16-0032 from EU Regional Development Fund.

Institutional Review Board Statement: Not applicable.

Informed Consent Statement: Not applicable.

Data Availability Statement: Data supporting the results presented can be provided upon request to the corresponding author.

Conflicts of Interest: The authors declare no conflict of interest. The funders had no role in the design of the study; in the collection, analyses, or interpretation of data; in the writing of the manuscript; or in the decision to publish the results.

proceedings

Abstract

A Brief Journey around the World of Heteracyclo[n]phanes: Synthesis and CO_2 Adsorption †

Jean-Manuel Raimundo

Département Matériaux Polytech Marseille, Aix Marseille Université, CNRS, AMUTech, CINaM 7325, 163 Avenue de Luminy, Case 913, 13288 Marseille, Cedex 09, France; jean-manuel.raimundo@univ-amu.fr

† Presented at the International Conference EcoBalt 2023 "Chemicals & Environment", Tallinn, Estonia, 9–11 October 2023.

Keywords: π-conjugated systems; macrocycles; CO_2 capture; advanced organic materials

The optoelectronic properties of π-conjugated chromophores have been widely used since the discovery of the conducting properties of synthetic organic polymers in the late 1970s. Indeed, due to their chemical structures, organic conducting polymers exhibit electrical properties that provide useful materials, such as batteries, light-emitting diodes, antistatic packaging and coatings, microelectronic devices, photovoltaic cells, corrosion inhibitors, sensors, etc. Among them, polyaniline and polysulfide have been extensively studied for this purpose due to their environmental stability, oxidation, or protonation-adjustable electrical properties and low-cost production.

Nevertheless, these structures are generally depicted as linear structures, and there is an interest in achieving hemicyclic or cyclic analogue structures to extend their potential for use as metal–gas sequestering agents and/or sensing materials, as well as new organic cyclic semiconductors exhibiting different properties from their linear counterparts. In this context, we have recently developed a simple strategy based on a nucleophilic aromatic substitution reaction to achieve such cyclic derivatives, which are parent compounds of calixarenes with heteroatoms on the bridge, affording the desired heteracalixarenes and mixed heteracalixarenes or the hemicyclic scaffold in high yields. Some examples and properties will be presented herein, as well as the on-going research toward identifying nanoarchitectures that can be integrated into devices such as sensors.

Citation: Raimundo, J.-M. A Brief Journey around the World of Heteracyclo[n]phanes: Synthesis and CO_2 Adsorption. *Proceedings* **2023**, *92*, 79. https://doi.org/10.3390/proceedings2023092079

Academic Editors: Monika Mortimer, Anne Kahru, Ivo Leito, Riin Rebane and Villem Aruoja

Published: 3 January 2024

Funding: This research received no external funding.

Institutional Review Board Statement: Not applicable.

Informed Consent Statement: Not applicable.

Data Availability Statement: Not applicable.

Conflicts of Interest: The authors declare no conflict of interest.

proceedings

MDPI

Abstract

Electrodeposited Cu Nanofoam Structures for Electrochemical CO_2 Reduction †

Birutė Serapinienė, Laima Gudavičiūtė and Rimantas Ramanauskas *

Center for Physical Sciences and Technology, Sauletekio 3, LT-10257 Vilnius, Lithuania; birute.serapiniene@ftmc.lt (B.S.); laima.gudaviciute@ftmc.lt (L.G.)

* Correspondence: rimantas.ramanauskas@ftmc.lt

† Presented at the International Conference EcoBalt 2023 "Chemicals & Environment", Tallinn, Estonia, 9–11 October 2023.

Keywords: 3D copper; electrochemically active real surface area; copper foam

An exclusive Cu surface feature has the ability to convert CO_2 into hydrocarbons with significant Faradaic efficiency [1]. The catalytic activity of metal is highly sensitive to electrolysis conditions including surface structure, morphology and real surface area (S_R). Simple polycrystalline Cu electrodes possess a rather small surface area; therefore, their efficiency is low. Three-dimensional nanoramified Cu electrodes or foams can be produced via metal electrodeposition with intensive hydrogen evolution [2]. An accurate estimation of the porous electrode S_R is of high importance, as precise knowledge of this parameter is crucial for comparison of the behaviour of various catalytic systems. The aim of this study was to evaluate the suitability of the known Cu real surface area determination methods [3] for their application for Cu 3D nanostructures. To reach this goal, the following design was created. The initial Cu electrode with a known S_R value was employed as a basis for Cu 3D structure electrodeposition from an acidic sulphate solution. Electrochemical methods employing the underpotential deposition of Tl and Pb, as well the double-layer capacitance measurements, applying cyclic voltammetry and electrochemical impedance spectroscopy were applied for Cu 3D structure S_R evaluation. The obtained results imply that non-porous Cu electrodes are not sensitive to the applied S_R determination method, while this parameter for Cu 3D structures depends significantly on the evaluation mode. The most reliable data for Cu foam characterization were obtained with double-layer capacity measurements, while all other applied methods yielded inaccurate results. The electrodeposited Cu 3D layer structure and hence S_R depend on the plating solution composition [4]. An attempt has been made to investigate the influence of HCl additives on deposited Cu foam S_R values. The obtained results indicate that the addition of HCl increases Cu S_R threefold.

Citation: Serapinienė, B.; Gudavičiūtė, L.; Ramanauskas, R. Electrodeposited Cu Nanofoam Structures for Electrochemical CO_2 Reduction. *Proceedings* **2023**, *92*, 4. https://doi.org/10.3390/proceedings2023092004

Monika Mortimer, Anne Kahru, Ivo Leito, Riin Rebane and Villem Aruoja

Published: 20 November 2023

Author Contributions: Conceptualization, R.R.; data curation, B.S., L.G. and R.R.; formal analysis, B.S., L.G. and R.R.; investigation, B.S. and L.G.; methodology, B.S., L.G. and R.R.; software, B.S. and L.G.; supervision, R.R.; writing—original draft, R.R.; writing—review and editing, R.R. All authors have read and agreed to the published version of the manuscript.

Funding: This research received no external funding.

Institutional Review Board Statement: Not applicable.

Informed Consent Statement: Not applicable.

Data Availability Statement: The data presented in this study are available on request from the corresponding author.

Conflicts of Interest: The authors declare no conflict of interest.

References

1. Xie, H.; Wang, T.; Liang, J.; Li, Q.; Sun, S. Cu-Based Nanocatalysts for Electrochemical Reduction of CO_2. *Nano Today* **2018**, *21*, 41–54. [CrossRef]
2. Xu, M.; Chen, Y.F.; Liang, J.Y.; Mo, D.C.; Lyu, S.S. Electrodeposition Patterned Copper Foam with Micro/Nanostructures for Reducing Supercooling in Water-Based Cool Storage Phase-Change Materials. *Appl. Sci.* **2020**, *10*, 4202. [CrossRef]
3. Trasatti, S.; Petrii, O.A. Real Surface Area Measurements in Electrochemistry. *Pure Appl. Chem.* **1991**, *63*, 711–734. [CrossRef]
4. Shin, H.C.; Liu, M. Copper Foam Structures with Highly Porous Nanostructured Walls. *Chem. Mater.* **2004**, *16*, 5460–5464. [CrossRef]

proceedings

Abstract

Nitrogen-Doped Reduced Graphene Oxide for Electrochemical Sensing Applications †

Justina Gaidukevič [1,2,*], Rūta Aukštakojytė [1], Mieczysław Kozłowski [3] and Rasa Pauliukaitė [2]

1 Institute of Chemistry, Faculty of Chemistry and Geosciences, Vilnius University, Naugarduko Str. 24, LT-03225 Vilnius, Lithuania; ruta.aukstakojyte@chgf.vu.lt
2 Department of Nanoengineering, Center for Physical Sciences and Technology, Savanoriu Ave. 231, LT-02300 Vilnius, Lithuania; rasa.pauliukaite@ftmc.lt
3 Faculty of Chemistry, Adam Mickiewicz University in Poznan, Uniwersytetu Poznanskiego 8, 61-614 Poznan, Poland; mieczyslaw.kozlowski@amu.edu.pl
* Correspondence: justina.gaidukevic@chf.vu.lt

† Presented at the International Conference EcoBalt 2023 "Chemicals & Environment", Tallinn, Estonia, 9–11 October 2023.

Keywords: nitrogen doping; graphene oxide; dopamine; electrochemical sensor

Graphene-based derivatives, such as graphene oxide (GO) and reduced graphene oxide (rGO), have gained significant attention in the field of electrochemical sensors [1]. These materials offer several advantages, including a time-efficient and cost-effective synthesis procedure and unique chemical, physical, and electronic properties [2]. rGO-based materials, in particular, possess a high surface area, chemical stability, and electrical conductivity [3]. However, despite these favorable characteristics, the development of rGO-based sensors with high sensitivity and rapid response times remains a challenge. One of the promising strategies to meet this challenge is the doping of rGO using heteroatoms. This approach involves introducing certain atoms (e.g., N, B, P, or S) into the graphene lattice, which can modify the structural and electrochemical rGO properties [4]. Therefore, this study focuses on the synthesis and structural characterization of N-doped rGO-based materials and their application to the electrochemical sensing of dopamine and H_2O_2. rGO modification with nitrogen species was achieved using two different synthesis approaches. To functionalize the rGO surface with a cationic Bismarck Brown dye, a hydrothermal synthesis method was employed. Also, the rGO surface was modified using gaseous ammonia at temperatures of 950 °C or 850 °C for 8 or 4 h, respectively. The obtained materials were characterized by different methods (XPS, BET, SEM, and Raman spectroscopy). Electrochemical measurements, such as cyclic voltammetry and chronoamperometry, were used to evaluate the obtained samples toward dopamine or H_2O_2 detection. The results demonstrated that various nitrogen species, including pyridinic-N, pyrrolic-N, and quaternary-N, were detected in the N-doped rGO. Moreover, it was observed that the amount and type of N-species introduced into the rGO surface contribute to the improved performance of the sensing platform, enabling the sensitive and selective detection of analytes.

Citation: Gaidukevič, J.; Aukštakojytė, R.; Kozłowski, M.; Pauliukaitė, R. Nitrogen-Doped Reduced Graphene Oxide for Electrochemical Sensing Applications. *Proceedings* **2023**, 92, 59. https://doi.org/10.3390/proceedings2023092059

Academic Editors: Monika Mortimer, Anne Kahru, Ivo Leito, Riin Rebane and Villem Aruoja

Published: 29 November 2023

Author Contributions: Conceptualization, J.G. and R.A.; methodology, J.G. and M.K.; investigation, J.G. and R.A.; resources, J.G.; data curation, J.G. and R.A.; writing—original draft preparation, J.G.; writing—review and editing, R.A., M.K. and R.P.; visualization, J.G.; supervision, R.P. and M.K.; funding acquisition, J.G. All authors have read and agreed to the published version of the manuscript.

Funding: This project received funding from the European Social Fund (project No 09.3.3-LMT-K-712-19-0050) under a grant agreement with the Research Council of Lithuania (LMTLT).

Institutional Review Board Statement: Not applicable.

Informed Consent Statement: Not applicable.

Data Availability Statement: The data presented in this study are available on request from the corresponding author.

Conflicts of Interest: The authors declare no conflict of interest.

References

1. Suvarnaphaet, P.; Pechprasarn, S. Graphene-Based Materials for Biosensors: A Review. *Sensors* **2017**, *17*, 2161. [CrossRef] [PubMed]
2. Jiříčková, A.; Jankovský, O.; Sofer, Z.; Sedmidubský, D. Synthesis and Applications of Graphene Oxide. *Materials* **2022**, *15*, 920. [CrossRef] [PubMed]
3. Ray, S.C. Chapter 2—Application and Uses of Graphene Oxide and Reduced Graphene Oxide. In *Applications of Graphene and Graphene-Oxide Based Nanomaterials*; Ray, S.C., Ed.; William Andrew Publishing: Oxford, UK, 2015; pp. 39–55.
4. Arvas, M.B.; Gürsu, H.; Gencten, M.; Sahin, Y. Preparation of different heteroatom doped graphene oxide based electrodes by electrochemical method and their supercapacitor applications. *J. Energy Storage* **2021**, *35*, 102328. [CrossRef]

proceedings MDPI

Abstract

Hemicucurbituril-Porphyrin Supramolecular Systems for Pollutant Sensing and Remediation †

Riina Aav *, Victor Borovkov, Dzmitry Kananovich, Nele Konrad, Tatsiana Jarg, Mari-Liis Brük and Marko Šakarašvili

Department of Chemistry and Biotechnology, Tallinn University of Technology, Akadeemia Tee 15, 12618 Tallinn, Estonia; victor.borovkov@taltech.ee (V.B.)
* Correspondence: riina.aav@taltech.ee
† Presented at the International Conference EcoBalt 2023 "Chemicals & Environment", Tallinn, Estonia, 9–11 October 2023.

Keywords: chiral receptor; pollutant; sensing; absorption; supramolecular systems

Citation: Aav, R.; Borovkov, V.; Kananovich, D.; Konrad, N.; Jarg, T.; Brük, M.-L.; Šakarašvili, M. Hemicucurbituril-Porphyrin Supramolecular Systems for Pollutant Sensing and Remediation. *Proceedings* **2023**, *92*, 87. https://doi.org/10.3390/proceedings2023092087

Academic Editors: Monika Mortimer, Anne Kahru, Ivo Leito, Riin Rebane and Villem Aruoja

Published: 30 January 2024

Hemicucurbiturils are members of the single-bridged cucurbituril family, formed through templated synthesis in a single step. This class of compounds is known for its ability to form inclusion complexes with electron-rich species. We have developed sustainable synthesis methods for chiral hemicucurbit[n]urils (where n = 6 or 8) and their derivatives [1–4]. Additionally, we have demonstrated that cyclohexanohemicucurbiturils can form external complexes with metalloporphyrins [5]. Porphyrins are well known for their optical and photochemical properties, which are extensively utilized in sensing and catalysis applications. Chirality sensing adds another viewpoint to sensing systems, and we have shown that toxic organocatalysts can be sensed via complex formation with Zn porphyrins [6]. By merging chiral hemicucurbiturils and metalloporphyrins via non-covalent interactions into a solid thin material, one can construct an enantioselective electronic nose and very selectively discriminate different analytes and their handedness [7]. In this conference, we will present our findings on how hemicucurbiturils and porphyrins, as supramolecular receptors, can be employed for optical and gravimetric sensing, as well as for the remediation of chemical pollutants.

Author Contributions: Conceptualization, R.A.; methodology, R.A., V.B. and D.K.; investigation, N.K., T.J., M.-L.B. and M.Š.; All authors have read and agreed to the published version of the manuscript.

Funding: This research was funded by the Estonian Research Council, grant number PRG399, and the European Union's H2020- FETOPEN, grant 828779 (INITIO).

Institutional Review Board Statement: Not applicable.

Informed Consent Statement: Not applicable.

Data Availability Statement: Not applicable.

Conflicts of Interest: The authors declare no conflicts of interest.

References

1. Aav, R.; Shmatova, E.; Reile, I.; Borissova, M.; Topic, F.; Rissanen, K. New Chiral Cyclohexylhemicucurbit[6]uril. *Org. Lett.* **2013**, *15*, 3786. [CrossRef] [PubMed]
2. Prigorchenko, E.; Öeren, M.; Kaabel, S.; Fomitšenko, M.; Reile, I.; Järving, I.; Tamm, T.; Topic, F.; Rissanen, K.; Aav, R. Template-controlled synthesis of chiral cyclohexylhemicucurbit[8]uril. *Chem. Commun.* **2015**, *51*, 10921. [CrossRef] [PubMed]
3. Kaabel, S.; Stein, R.S.; Fomitsenko, M.; Jarving, I.; Friscic, T.; Aav, R. Size-control by anion templating in mechanochemical synthesis of hemicucurbiturils in the solid state. *Angew. Chem. Int. Ed.* **2019**, *58*, 6230–6234. [CrossRef] [PubMed]
4. Dalidovich, T.; Mishra, K.A.; Shalima, T.; Kudrjašova, M.; Kananovich, D.G.; Aav, R. Mechanochemical Synthesis of Amides with Uronium-Based Coupling Reagents: A Method for Hexa-amidation of Biotin[6]uril. *ACS Sustain. Chem. Eng.* **2020**, *8*, 15703. [CrossRef]
5. Ustrnul, L.; Kaabel, S.; Burankova, T.; Martõnova, J.; Adamson, J.; Konrad, N.; Burk, P.; Borovkov, V.; Aav, R. Supramolecular chirogenesis in zinc porphyrins by enantiopure hemicucurbit[n]urils (n = 6, 8). *Chem. Commun.* **2019**, *55*, 14434. [CrossRef] [PubMed]
6. Konrad, N.; Horetski, M.; Sihtmäe, M.; Truong, K.N.; Osadchuk, I.; Burankova, T.; Kielmann, M.; Adamson, J.; Kahru, A.; Rissanen, K.; et al. Thiourea Organocatalysts as Emerging Chiral Pollutants: En Route to Porphyrin-Based (Chir)Optical Sensing. *Chemosensors* **2021**, *9*, 278. [CrossRef]
7. Magna, G.; Sakarasvili, M.; Stefanelli, M.; Giancane, G.; Bettini, S.; Valli, L.; Ustrnul, L.; Borovkov, V.; Aav, R.; Monti, D.; et al. Chiral Recognition by Supramolecular Porphyrin–Hemicucurbit[8]uril-Functionalized Gravimetric Sensors. *ACS Appl. Mater. Interfaces* **2023**, *15*, 30674. [CrossRef] [PubMed]

proceedings

Abstract

Anthropogenic Activities and Microbial Populations: War, Peace or Adaptation? †

Julija Armalytė and Eglė Lastauskienė *

Life Sciences Center, Vilnius University, Saulėtekio Av. 7, LT-10257 Vilnius, Lithuania; julija.armalyte@gf.vu.lt
* Correspondence: egle.lastauskiene@gf.vu.lt; Tel.:+370-5-239-8210

† Presented at the International Conference EcoBalt 2023 "Chemicals & Environment", Tallinn, Estonia, 9–11 October 2023.

Keywords: sediment metagenome; antibiotic resistance; resistome profiles

Aquaculture is one of the fastest-growing food sectors in Central, North, and Eastern Europe. Freshwater farming is changing the biodiversity of fishing ponds to fulfill industrial needs, and these changes can impact adjusting water bodies. Precautions should be taken to protect ecosystems and ensure that they are sustainable. The main objective of this research was to evaluate the influence of intensive fish farming on the condition of fishery ponds themselves and the surrounding water ecosystems, as well as to evaluate the possibility of transferring pollutants and antibiotic resistance genes to both environment and human hosts. Sediment samples and fish gut microbiome samples were collected during September 2019 and the Summer of 2020 in three locations in Lithuania: fishery ponds, Simnas Lake upstream from the fishery ponds, and Dusia Lake, which is downstream from the fishery ponds. Heavy metals and antibiotic residues were measured in the samples. Genomic DNA was isolated from the samples using the ZymoBIOMICSTM DNA Miniprep Kit according to the manufacturer's recommendations. The composition of the bacterial community was determined using next-generation sequencing (NGS) by scanning the amplicons of the bacterial 16S rRNA gene. The V3–V4 16S rRNA regions were chosen for sequencing because they are capable of detecting both bacterial and archaea taxons with a high resolution [1,2]. NGS was performed by Novogene Bioinformatics Technology Co., Ltd. (Beijing, China) on an Illumina paired-end platform to generate 250 base pairs (bps) length paired-end raw reads. None of the tested sediment samples showed significantly elevated heavy metal concentrations or substantial veterinary antibiotic pollution. From the antibiotic resistance genes tested, the presence of aminoglycoside and b-lactam resistance determinants, as well as the presence of integrons, could be of concern for the possibility of being transferred to humans. The microbiome beta-diversity analysis results clearly indicated the differences between the microbiota composition of all pond sediments and the entrance point, treated as a clean area.

Citation: Armalytė, J.; Lastauskienė, E. Anthropogenic Activities and Microbial Populations: War, Peace or Adaptation? *Proceedings* **2023**, *92*, 75. https://doi.org/10.3390/proceedings2023092075

Academic Editors: Monika Mortimer, Anne Kahru, Ivo Leito, Riin Rebane and Villem Aruoja

Published: 5 December 2023

Author Contributions: Conceptualization, E.L. and J.A.; methodology, E.L. and J.A.; resources, E.L. and J.A.; data curation, E.L. and J.A.; writing—original draft preparation, E.L.; writing—review and editing, E.L.; visualization, E.L. and J.A.; supervision, E.L.; project administration, E.L.; funding acquisition, E.L. and J.A. All authors have read and agreed to the published version of the manuscript.

Funding: This research was funded by a Grant (No. S-SIT-20-6) from the Research Council of Lithuania.

Institutional Review Board Statement: Ethical review and approval were waived for this study, due to the fact that the fish used in this study were catched according to the "Rules for AmateurFishing in Inland Waters", approved by the Ministry of Environment D1-435, 23 July 2019, Republic of Lithuania.

Informed Consent Statement: Not applicable.

Data Availability Statement: The datasets presented in this study can be found in online repositories. The names of the repository/repositories and accession number(s) can be found at: NCBI Bioproject PRJNA715198.

Acknowledgments: Authors would like to thank Modestas Ružauskas, Vaidotas Valskys, Vilmantas Gėgžna, and Justinas Kavoliūnas for the sample collection and data analysis, and Radvilė Drevinskaitė, Karina Kasperovičiūtė, Karolina Sabaitė, and Ieva Ščerba for excellent technical assistance.

Conflicts of Interest: The authors declare no conflict of interest.

References

1. Wear, E.K.; Wilbanks, E.G.; Nelson, C.E.; Carlson, C.A. Primer selection impacts specific population abundances but not community dynamics in a monthly time-series 16S rRNA gene amplicon analysis of coastal marine bacterioplankton. *Environ. Microbiol.* **2018**, *20*, 2709–2726. [CrossRef] [PubMed]
2. Bukin, Y.S.; Galachyants, Y.P.; Morozov, I.V.; Bukin, S.V.; Zakharenko, A.S.; Zemskaya, T.I. The effect of 16s rRNA region choice on bacterial community metabarcoding results. *Sci. Data* **2019**, *6*, 190007. [CrossRef] [PubMed]

proceedings

Abstract

Viral Effect on Carbon and Nitrogen Cycling in Bloom-Forming Cyanobacteria †

Sigitas Šulčius [1,*], Jolita Kuznecova [1], Adam Antosiak [2], Gediminas Alzbutas [1], Valiantsin Lukashevich [1] and Dariusz Dziga [2]

1 Laboratory of Algology and Microbial Ecology, Nature Research Centre, Akademijos st. 2, LT-08412 Vilnius, Lithuania; jol.kuznecova@gmail.com (J.K.); gediminas.alzbutas@gamtc.lt (G.A.); valiantsin.lukashevich@gamtc.lt (V.L.)

2 Laboratory of Metabolomics, Faculty of Biochemistry, Biophysics and Biotechnology, Jagiellonian University, Gronostajowa 7, 30387 Krakow, Poland; adam.antosiak@doctoral.uj.edu.pl (A.A.); dariusz.dziga@uj.edu.pl (D.D.)

* Correspondence: sigitas.sulcius@gamtc.lt; Tel.: +370-669-61007

† Presented at the International Conference EcoBalt 2023 "Chemicals & Environment", Tallinn, Estonia, 9–11 October 2023.

Keywords: *Aphanizomenon flos-aquae*; cyanophage vB_AphaS-CL131; harmful cyanobacterial blooms; metabolic reprogramming

Viruses can significantly influence the biogeochemical cycling of major nutrients through the infection and lysis of cyanobacteria, a globally important primary producer [1]. However, surprisingly little attention has been given to understanding how viruses alter the metabolism of carbon (C) and nitrogen (N) in bloom-forming cyanobacteria, distributed worldwide in fresh and brackish water ecosystems. Moreover, there is a lack of information about how co-occurring microbial communities respond to the lysis of these primary producers. Therefore, we employed an ecologically relevant filamentous diazotrophic cyanobacteria *Aphanizomenon flos-aquae* [2] and *Nodularia spumigena* [3], and their lytic cyanophages [4,5], as host–virus model systems in combination with a series of incubation experiments, to investigate the effect of viral infection and lysis on photosynthetic activity, nitrogen assimilation and enrichment rates, expression levels of genes involved in photosynthesis, and carbon and nitrogen metabolism, as well as on the concentration of some central and secondary cellular metabolites. In addition, we analyzed the variation in the composition of associated bacterial assemblages in response to viral additions and in relation to uninfected cyanobacterial cultures throughout their cultivation periods. We found that the effect of cyanophages on carbon and nitrogen cycling and cellular metabolism was significant yet varied widely depending on the stage of the infection process (e.g., cyanophage adsorption vs. DNA replication vs. release), and the state of the host culture (culture undergoing infection/lysis vs. recovering culture). Our observations suggest that cyanobacteria underwent a physiological state shift towards lower efficiency carbon and energy cycling, as well as to the reduced nitrogen transport from heterocytes (N-fixing cells) to vegetative cells [6,7]. The lysis of cyanobacterial cells was associated with a release of ammonium and other compounds that promoted changes in co-occurring microbes. The shift in the associated bacterial community was related to the infection rate and increased with higher initial cyanophage density. On the contrary, the initial infection rate, although it affected the timing, had no effect on the magnitude of net population loss or changes in population structure. Our observations indicate that cyanophage infection and lysis have implications across multiple levels of ecological organization, from cell to population and to the entire community [5,6].

Citation: Šulčius, S.; Kuznecova, J.; Antosiak, A.; Alzbutas, G.; Lukashevich, V.; Dziga, D. Viral Effect on Carbon and Nitrogen Cycling in Bloom-Forming Cyanobacteria. *Proceedings* **2023**, *92*, 36. https://doi.org/10.3390/proceedings2023092036

Academic Editors: Monika Mortimer, Anne Kahru, Ivo Leito, Riin Rebane and Villem Aruoja

Published: 23 November 2023

Author Contributions: Conceptualization, S.Š.; methodology, S.Š. and D.D.; formal analysis, S.Š., J.K., A.A., G.A., V.L. and D.D.; investigation, J.K., A.A. and V.L.; resources, S.Š. and D.D.; data curation, S.Š., A.A., G.A. and D.D.; writing—original draft preparation, S.Š.; writing—review and editing, S.Š.; visualization, S.Š., G.A. and A.A.; supervision, S.Š. and D.D.; project administration, S.Š. and D.D.; funding acquisition, S.Š. and D.D. All authors have read and agreed to the published version of the manuscript.

Funding: This research was funded by the Research Council of Lithuania, grant number S-LL-21-10, to S.Š., and by the Nature Research Centre, through open access to the research infrastructure of the Nature Research Centre under the Lithuanian open access network initiative. J.K. was also supported by Deutsche Bundesstiftung Umwelt (DBU), scholarship number 30018/772. D.D. and A.A. were supported by the National Science Centre of Poland, project number 2020/38/L/NZ9/00135.

Institutional Review Board Statement: Not applicable.

Informed Consent Statement: Not applicable.

Data Availability Statement: Data supporting this study are openly available from public repositories (https://doi.org/10.1128/AEM.01311-18, https://doi.org/10.1016/j.hal.2018.07.004, https://doi.org/10.3389/fmicb.2020.02010, https://doi.org/10.1016/j.hal.2022.102215). Part of the data is not publicly available due to restrictions of the Research Council of Lithuania and the National Science Centre of Poland due restrictions for ongoing research projects. All data presented in this study are available from the corresponding author upon reasonable request.

Acknowledgments: The authors are grateful to Klaus Jürgens, Maren Voss, and their lab members from the Leibniz Institute for Baltic Sea Research (Warnemünde, Germany) for providing support during incubation experiments and their help with NanoSIMS analysis. We also thank Jūratė Kasparovičienė from the Nature Research Centre (Vilnius, Lithuania) for the enumeration of cyanobacteria cells, P. Malec from Jagiellonian University (Krakow, Poland) for the help with photosynthetic activity measurements, and H. Mazur-Marzec from University of Gdańsk for the analysis of non-ribosomal peptides, as well as for their comments and suggestions during the preparation of this abstract.

Conflicts of Interest: The authors declare no conflict of interest.

References

1. Jover, L.F.; Effler, T.C.; Buchan, A.; Wilhelm, S.W.; Weitz, J.S. The elemental composition of virus particles: Implications for marine biogeochemical cycles. *Nat. Rev. Microbiol.* **2014**, *12*, 519–528. [CrossRef] [PubMed]
2. Cirés, S.; Ballot, A. A review of the phylogeny, ecology and toxin production of bloom-forming *Aphanizomenon* spp. and related species within the *Nostocales* (cyanobacteria). *Harmful Algae* **2016**, *54*, 21–43. [CrossRef] [PubMed]
3. Bertos-Fortis, M.; Farnelid, H.M.; Lindh, M.V.; Casini, M.; Andersson, A.; Pinhassi, J.; Legrand, C. Unscrambling cyanobacteria community dynamics related to environmental factors. *Front. Microbiol.* **2016**, *7*, 625. [CrossRef] [PubMed]
4. Šulčius, S.; Šimoliūnas, E.; Alzbutas, G.; Gasiūnas, G.; Jauniškis, V.; Kuznecova, J.; Miettinen, S.; Nilsson, E.; Meškys, R.; Roine, E.; et al. Genomic characterization of cyanophage vB_AphaS-CL131 infecting filamentous diazotrophic cyanobacterium *Aphanizomenon flos-aquae* reveals novel insights into virus-bacterium interactions. *Appl. Environ. Microbiol.* **2019**, *85*, e01311–e01318. [CrossRef] [PubMed]
5. Šulčius, S.; Mazur-Marzec, H.; Vitonytė, I.; Kvederavičiūtė, K.; Kuznecova, J.; Šimoliūnas, E.; Holmfeldt, K. Insights into cyanophage-mediated dynamics of nodularin and other non-ribosomal peptides in *Nodularia spumigena*. *Harmful Algae* **2018**, *78*, 69–74. [CrossRef] [PubMed]
6. Kuznecova, J.; Šulčius, S.; Vogts, A.; Voss, M.; Jürgens, K.; Šimoliūnas, E. Nitrogen flow in diazotrophic cyanobacterium *Aphanizomenon flos-aquae* is altered by cyanophage infection. *Front. Microbiol.* **2020**, *11*, 2010. [CrossRef] [PubMed]
7. Antosiak, A.; Šulčius, S.; Malec, P.; Tokodi, N.; Łobodzińska, A.; Dziga, D. Cyanophage infections reduce photosynthetic activity and expression of CO_2 fixation genes in the freshwater bloom-forming cyanobacterium *Aphanizomenon flos-aquae*. *Harmful Algae* **2022**, *116*, 102215. [CrossRef] [PubMed]

proceedings

Abstract

Use of Neoteric Solvents in Biomass Treatment †

Merike Vaher [1], Piret Saar-Reismaa [1,2], Maria Kuhtinskaja [1] and Mihkel Koel [1,*]

[1] Institute of Chemistry and Biotechnology, Tallinn University of Technology, Tallinn 12918, Estonia; merike.vaher@taltech.ee (M.V.); pirsaar@gmail.com (P.S.-R.); maria.kuhtinskaja@taltech.ee (M.K.)
[2] HAN BioCentre, Nijmegen 6503, The Netherlands
* Correspondence: mihkel.koel@taltech.ee; Tel.: +372-5215676

† Presented at the International Conference EcoBalt 2023 "Chemicals & Environment", Tallinn, Estonia, 9–11 October 2023.

Keywords: supercritical fluids; ionic liquids; deep eutectic solvents; extraction; chromatography; essential oils; bioactive compounds

Research on the use of neoteric solvents has been motivated by an increase in pollution-controlled legislation and more regulations of common solvents, especially related to the treatment of biomass. The most widely used neoteric solvent in biomass treatment, and especially for the extraction of essential oils and other bioactive compounds from plants, is supercritical carbon dioxide. A great reason to use CO_2 in its supercritical condition is to expand the spectrum of solvent solubility, polarity, and volatility [1].

Ionic liquids (ILs) and the deep eutectic solvents (DESs) are very attractive for extraction and separation sciences as replacements for volatile organic solvents. Unfortunately, traditional ILs, despite their high chemical flexibility and non-volatility, are not green solvents.

DESs represent another type of tailor-made solvent that use both hydrogen bond donor (HBD) and hydrogen bond acceptor (HBA) compounds, available from renewable resources, resulting in natural deep eutectic solvents (NADESs) which have preferable intrinsic characteristics, low cost, easy preparation, and low toxicity. The HBDs and HBAs of a DES dictate the solvent properties that have a direct impact on the extraction efficiency. The role of water in DESs' composition is to alter the pH, viscosity, and polarity, which results in a significant increase in extraction efficiency.

There are studies that have shown the excellent performances of DESs in the extraction of biomolecules (such as polyphenols, iridoids, and alkaloids), providing comparable or even higher efficiency than conventional solvents [2,3].

A variety of analytical methods could be applied for the characterization of obtained extracts, including HPLC-DAD-MS, which is well suited for analysing the most complicated samples.

Citation: Vaher, M.; Saar-Reismaa, P.; Kuhtinskaja, M.; Koel, M. Use of Neoteric Solvents in Biomass Treatment. *Proceedings* **2023**, *92*, 39. https://doi.org/10.3390/proceedings2023092039

Academic Editors: Monika Mortimer, Anne Kahru, Ivo Leito, Riin Rebane and Villem Aruoja

Published: 23 November 2023

Author Contributions: Conceptualization, M.K. (Mihkel Koel) and M.V.; methodology, M.V. and M.K. (Mihkel Koel); validation, M.K. (Mihkel Koel), M.K. (Maria Kuhtinskaja) and M.V.; investigation, M.V., P.S.-R. and M.K. (Maria Kuhtinskaja); data curation, M.V.; writing—original draft preparation, M.K. (Mihkel Koel); writing—review and editing, M.V., P.S.-R. and M.K. (Maria Kuhtinskaja); project administration, M.V.; funding acquisition, M.V. All authors have read and agreed to the published version of the manuscript.

Funding: This research was funded by the Estonian Research Council (Grant IUT33-20) and the R&D project SS22004 "Evaluation of antioxidant and antibacterial activity of plant extracts", funded by the Tallinn University of Technology.

Institutional Review Board Statement: Not applicable.

Informed Consent Statement: Not applicable.

Data Availability Statement: Data will be available on request from authors.

Acknowledgments: The Estonian Center of Analytical Chemistry (ECAC) is acknowledged for advice on methodology.

Conflicts of Interest: The authors declare no conflict of interest.

References

1. Gretšušnikova, T.; Järvan, K.; Orav, A.; Koel, M. Comparative analysis of the composition of the essential oil from the shoots, leaves and stems the wild *Ledum palustre* L. from Estonia. *Procedia Chem.* **2010**, *2*, 168–173. [CrossRef]
2. Koel, M.; Kuhtinskaja, M.; Vaher, M. Extraction of bioactive compounds from *Catharanthus roseus* and *Vinca minor*. *Sep. Purif. Technol.* **2020**, *252*, 117438. [CrossRef]
3. Saar-Reismaa, P.; Koel, M.; Tarto, R.; Vaher, M. Extraction of bioactive compounds from *Dipsacus fullonum* leaves using deep eutectic solvents. *J. Chromatogr. A* **2022**, *1677*, 463330. [CrossRef] [PubMed]

proceedings

Abstract

Sustainable Chemistry through Catalysis and Process Intensification †

Henrik Grénman

Johan Gadolin Process Chemistry Centre, Laboratory of Industrial Chemistry and Reaction Engineering, Åbo Akademi University, 20500 Turku, Finland; henrik.grenman@abo.fi; Tel.: +358-505764744

† Presented at the International Conference EcoBalt 2023 "Chemicals & Environment", Tallinn, Estonia, 9–11 October 2023.

Keywords: heterogeneous catalysis; process intensification; effluents treatment; liquid and gas phase processing

The shift away from fossil resources is revolutionizing our industrial carbon sources, and the developments in legislation demand increased overall efficiency in processes and emission abatement. Catalysis plays a key role in enabling the green transition in the chemical process industry and environmental protection. They also act as a bridge between chemical reactions and reaction mechanisms and moving from the molecular to the process scale. Besides enhancing reaction rates, increasing selectivity plays a key role, and both of these factors are tightly linked also to the process design and optimization for which modern process intensification provides good tools. The current presentation displays three examples of combining heterogeneous catalysis with process intensification for wastewater treatment and the direct conversion of CO_2 recently studied by our research group. The wastewater treatment includes removing hemicelluloses from dilute biorefinery effluents with the help of catalytic aqueous-phase reforming in a continuous reactor [1–4]. The second case focuses on the removal of pharmaceuticals from communal wastewaters by combining ozonation with heterogeneous catalysis in a semi-batch reactor operating at ambient pressure [5–7]. In the case focusing on gas-phase processing, CO_2 is converted to renewable natural gas utilizing a bi-functional catalytic material in a periodically operating continuous reactor concept [8–12]. Chromatography was used as the main analysis method in all of the experiments. High yields and good selectivity were obtained in all of the cases, and the next steps are related to process optimization, stability testing, and preparative studies for scale-up studies. The obtained results display significant potential for green process technology and process efficiency by combining catalyst development with process design to be able to efficiently utilize effluent streams and minimize the effects on the environment.

Citation: Grénman, H. Sustainable Chemistry through Catalysis and Process Intensification. *Proceedings* **2023**, *92*, 76. https://doi.org/10.3390/proceedings2023092076

Academic Editors: Monika Mortimer, Anne Kahru, Ivo Leito, Riin Rebane and Villem Aruoja

Published: 13 December 2023

Funding: This research was financially supported by the Academy of Finland, Business Finland, and Tekniikan Edistämissäätiö.

Institutional Review Board Statement: Not applicable.

Informed Consent Statement: Not applicable.

Data Availability Statement: Data is available via the cited articles and from the corresponding author.

Conflicts of Interest: The author declares no conflict of interest.

References

1. Aho, A.; Alvear, M.; Ahola, J.; Kangas, J.; Tanskanen, J.; Simakova, I.; Santos, J.; Eränen, K.; Salmi, T.; Murzin, D.Y.; et al. Aqueous phase reforming of birch and pine hemicellulose hydrolysates. *Bioresour. Technol.* **2022**, *348*, 126809. [CrossRef] [PubMed]
2. Aho, A.; Õna, J.P.; Rosales, C.; Eränen, K.; Salmi, T.; Murzin, D.Y.; Grénman, H. Biohydrogen from dilute side streams—Influence of reaction conditions on the conversion and selectivity in aqueous phase reforming of xylitol. *Biomass Bioenergy* **2020**, *138*, 105590. [CrossRef]
3. Alvear, M.; Aho, A.; Simakova, I.; Grénman, H.; Salmi, T.; Murzin, D.Y. Aqueous phase reforming of alcohols over a bimetallic Pt-Pd catalyst in the presence of formic acid. *Chem. Eng. J.* **2020**, *398*, 15–125541. [CrossRef]
4. Alvear, M.; Aho, A.; Simakova, I.; Grénman, H.; Salmi, T.; Murzin, D.Y. Aqueous phase reforming of xylitol and xylose in the presence of formic acid. *Catal. Sci. Technol.* **2020**, *10*, 5245–5255. [CrossRef]
5. Saeid, S.; Tolvanen, P.; Kumar, N.; Eränen, K.; Peltonen, J.; Peurla, M.; Mikkola, J.P.; Franz, A.; Salmi, T. Advanced oxidation process for the removal of ibuprofen from aqueous solution: A non-catalytic and catalytic ozonation study in a semi-batch reactor. *Appl. Catal. B Environ.* **2018**, *230*, 77–90. [CrossRef]
6. Saeid, S.; Kråkström, M.; Tolvanen, P.; Kumar, N.; Eränen, K.; Mikkola, J.P.; Kronberg, L.; Eklund, P.; Peurla, M.; Aho, A.; et al. Advanced oxidation process for degradation of carbamazepine from aqueous solution: Influence of metal modified microporous, mesoporous catalysts on the ozonation process. *Catalysts* **2020**, *10*, 90. [CrossRef]
7. Kråkström, M.; Saeid, S.; Tolvanen, P.; Salmi, T.; Kronberg, L.; Eklund, P. Catalytic ozonation of the antibiotic sulfadiazine: Reaction kinetics and transformation mechanisms. *Chemosphere* **2020**, *247*, 125853. [CrossRef] [PubMed]
8. Wei, L.; Haije, W.; Grénman, H.; de Jong, W. Sorption enhanced catalysis for CO_2 hydrogenation towards fuels and chemicals with focus on methanation. In *Heterogeneous Catalysis*; Elsevier: Amsterdam, The Netherlands, 2022; pp. 95–119.
9. Wei, L.; Azad, H.; Hamza; Haije, W.; Grénman, H.; de Jong, W. Pure methane from CO_2 hydrogenation using a sorption enhanced process with Catalyst/Zeolite bifunctional materials. *Appl. Catal. B Environ.* **2021**, *297*, 120399. [CrossRef]
10. Wei, L.; Grénman, H.; Haije, W.; Kumar, N.; Aho, A.; Eränen, K.; Wei, L.; de Jong, W. Sub-nanometer ceria-promoted Ni 13X zeolite catalyst for CO_2 methanation. *Appl. Catal. A Gen.* **2021**, *612*, 118012. [CrossRef]
11. Wei, L.; Haije, W.; Kumar, N.; Peltonen, J.; Peurla, M.; de Jong, W.; Grénman, H. The influence of nickel precursors on the properties and performance of Ni impregnated zeolite 5A and 13X supported catalysts in CO_2 methanation. *Catal. Today* **2021**, *362*, 35–46. [CrossRef]
12. Wei, L.; Haije, W.; Kumar, N.; de Jong, W.; Grénman, H. Can bi-functional nickel modified 13X and 5A zeolite catalysts for CO_2 methanation be improved by introducing ruthenium? *Mol. Catal. A* **2020**, *494*, 111115. [CrossRef]

proceedings

Abstract

Microplastics and Associated Microorganisms in the Sea Sediment of the Sentina Regional Natural Reserve (Central Adriatic Sea, Italy) †

Angela Piersanti [1], Anna-Rita Attili [1], Monika Mortimer [2], Signe Vahur [3], Francesco Alessandro Palermo [1], Paolo Cocci [1] and Cristina Miceli [1,*]

1 School of Biosciences and Veterinary Medicine, Cell Biology and Biotechnology, University of Camerino, 62032 Camerino, Italy; angela.piersanti@unicam.it (A.P.); annarita.attili@unicam.it (A.-R.A.); francesco.palermo@unicam.it (F.A.P.); paolo.cocci@unicam.it (P.C.)
2 Institute of Environmental and Health Sciences, China Jiliang University, Hangzhou 310018, China; monika.mortimer@kbfi.ee
3 Institute of Chemistry, Faculty of Science and Technology, University of Tartu, 50411 Tartu, Estonia; signe.vahur@ut.ee
* Correspondence: cristina.miceli@unicam.it
† Presented at the International Conference EcoBalt 2023 "Chemicals & Environment", Tallinn, Estonia, 9–11 October 2023.

Keywords: microplastics; antibiotic resistance; metagenomics; bioremediation; microplastics isolation; sediment analysis

(1) Background: The large dispersion of microplastics (MPs) in the marine environment has effects on the health of living organisms [1–3]. The aim of this study was to identify MPs and their associated microorganisms in Adriatic Sea sediments and to evaluate the antibiotic susceptibility patterns of the microbial communities. (2) Methods: A beach transect, parallel to the shoreline where the waves break, was identified for the samplings. A protocol to perform MP isolation from sandy sediments at different seasons, suitable for biological sample upkeeping, and based on plastic floating in high salinity water, was optimized. From floating MPs, aerobic and anaerobic cultivable microorganisms were isolated and total DNA extraction was performed for the shotgun metagenomic analysis. Susceptibility to a panel of 14 antibiotics, belonging to 12 different categories, was assessed [4,5]. Chemical characteristics of the isolated MPs were analysed using a Thermo Nicolet 6700 FT-IR Spectrometer with "Smart Orbit" diamond micro-ATR accessory and Thermo Nicolet iN10 MX FT-IR microscope. (3) Results: Via chemical analysis, polypropylene microplastics were estimated in the highest percentage, followed by polyethylene, poly-methyl acrylate, and poly-vinyl chloride. Metagenomics data revealed differences in bacterial abundances during seasons and in floating MPs with respect to total sand. The differential gene analysis showed specific metabolic pathways in MP-associated microorganisms, including antibiotic resistance. Via microbial cultivation and MALDI-TOF MS identification, bacteria that are promising for plastic degradation, such as *Lysinobacillus fusiformis*, *Exiguobacterium* sp., and *Pseudomonas oleovorans*, were also found, as well as potential pathogens, like *Clostridium septicum*, *Clostridium novyi*, and *Shewanella putrefaciens*. Only 17.2% were found to be susceptible to all the tested antibiotics. High percentages of resistance were observed for penicillins (85.7%), monobactams (80.9%), and tetracyclines (64.3%). (4) Conclusions: MPs work as a vehicle for potential pathogens and antibiotic-resistant microorganisms in the Central Adriatic Sea.

Citation: Piersanti, A.; Attili, A.-R.; Mortimer, M.; Vahur, S.; Palermo, F.A.; Cocci, P.; Miceli, C. Microplastics and Associated Microorganisms in the Sea Sediment of the Sentina Regional Natural Reserve (Central Adriatic Sea, Italy). *Proceedings* **2023**, 92, 78. https://doi.org/10.3390/proceedings2023092078

Academic Editors: Anne Kahru, Ivo Leito, Riin Rebane and Villem Aruoja

Published: 28 December 2023

Author Contributions: Conceptualization, C.M., F.A.P. and M.M.; methodology, A.P., A.-R.A., M.M. and S.V.; formal analysis and data curation, A.P., A.-R.A. and P.C.; writing—review and editing, A.P., M.M. and C.M. All authors have read and agreed to the published version of the manuscript.

Funding: This research was funded by the Fondazione CRUI GoForIT project to support A.P.'s fellowship and Far BVI000068 to C.M.

Institutional Review Board Statement: Not applicable.

Informed Consent Statement: Not applicable.

Data Availability Statement: The metagenomic data (sequenced DNA reads) mentioned in this study are available on request from the corresponding author. The data are not publicly available yet, since they need a deeper investigation.

Conflicts of Interest: The authors declare no conflicts of interest.

References

1. Ogonowski, M.; Motiei, A.; Ininbergs, K.; Hell, E.; Gerdes, Z.; Udekwu, K.I.; Bacsik, Z.; Gorokhova, E. Evidence for selective bacterial community structuring on microplastics. *Environ. Microbiol.* **2018**, *20*, 2796–2808. [CrossRef] [PubMed]
2. Lu, L.; Luo, T.; Zhao, Y.; Cai, C.; Fu, Z.; Jin, Y. Interaction between microplastics and microorganism as well as gut microbiota: A consideration on environmental animal and human health. *Sci. Total Environ.* **2019**, *667*, 94–100. [CrossRef]
3. Silva, A.B.; Bastos, A.S.; Justino, C.I.; da Costa, J.P.; Duarte, A.C.; Rocha-Santos, T.A. Microplastics in the environment: Challenges in analytical chemistry—A review. *Anal. Chim. Acta* **2018**, *1017*, 1–19. [CrossRef] [PubMed]
4. Zhang, Y.; Lu, J.; Wu, J.; Wang, J.; Luo, Y. Potential risks of microplastics combined with superbugs: Enrichment of antibiotic resistant bacteria on the surface of microplastics in mariculture system. *Ecotoxicol. Environ. Saf.* **2022**, *187*, 109852. [CrossRef]
5. Su, Y.; Zhang, Z.; Zhu, J.; Shi, J.; Wei, H.; Xie, B.; Shi, H. Microplastics act as vectors for antibiotic resistance genes in landfill leachate: The enhanced roles of the long-term aging process. *Environ. Pollut.* **2021**, *270*, 116278. [CrossRef] [PubMed]

proceedings

MDPI

Abstract

Novel Plasticizers Are Emerging Contaminants †

Margit Heinlaan *, Heiki Vija and Irina Blinova

Laboratory of Environmental Toxicology, National Institute of Chemical Physics and Biophysics, Akadeemia tee 23, 12618 Tallinn, Estonia; heiki.vija@kbfi.ee (H.V.); irina.blinova@kbfi.ee (I.B.)

* Correspondence: margit.heinlaan@kbfi.ee

† Presented at the International Conference EcoBalt 2023 "Chemicals & Environment", Tallinn, Estonia, 9–11 October 2023.

Keywords: phthalates; wastewater sludge; DEHP; DPHP; DEHT; DINCH; plastic

In recent years, plastic use and pollution have gained a lot of attention. Plastic particles have been identified from the Mariana Trench [1] to high mountains [2]. However, the impact of plastic goes beyond that of particles [3]. Plastics are chemically very diverse. In addition to the principal polymeric component, plastics contain over 10,000 different additives [4] that may be non-intentional (e.g., production impurities) or intentional (functional) for obtaining certain properties. Flexibility is an important property for which plasticizers, a group of functional additives, are used. Until recently, phthalates have been the most widely used plasticizers, but advanced knowledge on the hazardous properties of several commonly used phthalate plasticizers has led to their strict regulation in the European Economic Area since 2020. This has created the need for alternative plasticizers, especially in sensitive human applications. The plastic industry has started producing and using novel non-phthalate plasticizers and high-molecular-weight phthalate plasticizers to the extent that they have become emerging contaminants [5,6]. Due to their hydrophobicity, plasticizers sorb to particulate matter and tend to accumulate in, e.g., wastewater treatment sludge. In the current research, sludge from wastewater treatment plants was sampled from all over Estonia (20 samples) to analyse the occurrence of selected novel plasticizers in the Estonian environment. Samples were analysed using gas chromatography–mass spectrometry (GC-MS). The first results showed that the most widely used DEHT (Bis(2-ethylhexylterephthalate), DPHP (Di(2-propylheptyl)phthalate), DINCH (1,2-cyclohexanedicarboxylic acid diisonyl ester) were above the GC-MS quantification limits in the majority of the analysed samples, and their levels were comparable to those of the regulated phthalate plasticizer DEHP (Bis(2-ethylhexyl) phthalate).

Citation: Heinlaan, M.; Vija, H.; Blinova, I. Novel Plasticizers Are Emerging Contaminants. *Proceedings* **2023**, *92*, 61. https://doi.org/10.3390/proceedings2023092061

Academic Editors: Monika Mortimer, Anne Kahru, Ivo Leito, Riin Rebane and Villem Aruoja

Published: 29 November 2023

Author Contributions: Conceptualization, M.H.; methodology, H.V.; software, H.V.; investigation, M.H. and H.V.; resources, M.H.; data curation, M.H. and H.V.; writing—original draft preparation, M.H.; writing—review and editing, I.B.; project administration, M.H.; funding acquisition, M.H. All authors have read and agreed to the published version of the manuscript.

Funding: This research was funded by the Estonian Research Council grant PRG1427.

Institutional Review Board Statement: Not applicable.

Informed Consent Statement: Not applicable.

Data Availability Statement: Not applicable.

Acknowledgments: The wastewater treatment plants leaders' willingness to cooperate is highly appreciated. Sampling was performed by Janek Reinik. The authors wish to thank Regine Nagorka and Jan Koschorreck from the German Environment Agency for advising on the plasticizer GC-MS protocol's optimization.

Conflicts of Interest: The authors declare no conflict of interest. The funders had no role in the design of the study; in the collection, analyses, or interpretation of data; in the writing of the manuscript; or in the decision to publish the results.

References

1. Jamieson, A.J.; Brooks, L.S.R.; Reid, W.D.K.; Piertney, S.B.; Narayanaswamy, B.E.; Linley, T.D. Microplastics and synthetic particles ingested by deep-sea amphipods in six of the deepest marine ecosystems on Earth. *Roy. Soc. Open sci.* **2019**, *6*, 180667. [CrossRef] [PubMed]
2. Tsering, T.; Sillanpää, M.; Viitala, M.; Reinikainen, S.-P. Variation of microplastics in the shore sediment of high-altitude lakes of the Indian Himalaya using different pretreatment methods. *Sci. Total Environ.* **2022**, *849*, 157870. [CrossRef] [PubMed]
3. Zimmermann, L.; Dierkes, G.; Ternes, T.A.; Völker, C.; Wagner, M. Benchmarking the in Vitro Toxicity and Chemical Composition of Plastic Consumer Products. *Environ. Sci. Technol.* **2019**, *53*, 11467–11477. [CrossRef] [PubMed]
4. Wiesinger, H.; Wang, Z.; Hellweg, S. Deep Dive into Plastic Monomers, Additives, and Processing Aids. *Environ. Sci. Technol.* **2021**, *55*, 9339–9351. [CrossRef] [PubMed]
5. Lemke, N.; Murawski, A.; Lange, R.; Weber, T.; Apel, P.; Dębiak, M.; Koch, H.M.; Kolossa-Gehring, M. Substitutes mimic the exposure behaviour of REACH regulated phthalates—A review of the German HBM system on the example of plasticizers. *Int. J. Hyg. Environ. Health* **2021**, *236*, 113780. [CrossRef] [PubMed]
6. Nagorka, R.; Birmili, W.; Schulze, J.; Koschorreck, J. Diverging trends of plasticizers (phthalates and non-phthalates) in indoor and freshwater environments—why? *Environ. Sci. Eur.* **2022**, *34*, 46. [CrossRef]

proceedings

MDPI

Abstract

Droplet-Based Technology for Studying the Phenotypic Effect of Microplastics on Antimicrobial Resistance †

Simona Bartkova, Fenella Lucia Sulp, Immanuel Sanka, Pille Pata and Ott Scheler *

Laboratory of Microfluidics, Department of Chemistry and Biotechnology, School of Science, Tallinn University of Technology, Akadeemia tee 15, 12618 Tallinn, Estonia; simona.bartkova@taltech.ee (S.B.); fenella.sulp@taltech.ee (F.L.S.); immanuel.sanka@taltech.ee (I.S.); pille.pata@taltech.ee (P.P.)

* Correspondence: ott.scheler@taltech.ee

† Presented at the International Conference EcoBalt 2023 "Chemicals & Environment", Tallinn, Estonia, 9–11 October 2023.

Keywords: antimicrobial resistance; microplastic pollution; droplet microfluidics; droplet-based technology; biofilm formation; plastisphere; phenotypic analysis; single cell incubation; Fluorescence microscopy

Plastic pollution is a global emergency [1,2]. One key problem is that microplastics (MPs) (1 µm–5 mm) [3] and nanoplastics (NPs) (≤1 µm) [4] enhance the already severe threat of antimicrobial resistance (AMR) by providing a micro-environment termed the "plastisphere" for bacteria to form biofilms [5,6]. However, exact knowledge of the severity of the plastisphere and its impact on AMR is currently still scarce [7]. Here, we show how droplet-based technology can be used to study the potential phenotypic effect of MPs on AMR. For this we used (i) polydisperse water-in-oil droplets generated via vortexing, (ii) GFB-labelled *Escherichia coli* JEK 1036 as our study object, (iii) cefotaxime as the test antibiotic, and (iv) 10 µm carboxylated polystyrene microspheres (PS). In parallel, we encapsulated single cells of *E. coli* into droplets with different concentrations of cefotaxime and with or without PS. After overnight incubation at 37 °C, we imaged droplets as a monolayer via confocal microscopy and analyzed droplets via Software Ilastik [8], CellProfiler^{TM} [9] and EasyFlow [10]. Our results show that *E. coli*'s minimal inhibitory concentration (MIC) shifts slightly towards a higher cefotaxime concentration when PS is present in droplets. Image analysis of *E. coli* growth patterns in individual droplets illustrates that *E. coli* tends to clump together in droplets with PS, versus exhibiting an evenly distributed growth pattern in droplets without PS. In conclusion, we see that PS in droplets might enhance the MIC of *E. coli* resistance against cefotaxime. This possible enhanced resistance may be related to the observed tendency for clumping (indication of biofilm formation) of *E. coli* when PS is present. Droplet-based technology is thus a suitable tool for studying the phenotypic effect of MPs on AMR. Further experiments with different antibiotics and MP types and sizes will shed more light on the interesting and worrying tendency of MPs to potentially enhance AMR that was found in this study.

Citation: Bartkova, S.; Sulp, F.L.; Sanka, I.; Pata, P.; Scheler, O. Droplet-Based Technology for Studying the Phenotypic Effect of Microplastics on Antimicrobial Resistance. *Proceedings* **2023**, *92*, 41. https://doi.org/10.3390/proceedings2023092041

Academic Editors: Monika Mortimer, Anne Kahru, Ivo Leito, Riin Rebane and Villem Aruoja

Published: 24 November 2023

Author Contributions: Conceptualization, S.B. and O.S.; methodology, S.B., F.L.S. and P.P.; software, S.B. and I.S.; validation, S.B. and F.L.S.; formal analysis, S.B., F.L.S. and O.S.; investigation, S.B. and O.S.; resources, O.S.; data curation, S.B. and I.S.; writing—original draft preparation, S.B.; writing—review and editing, S.B. and O.S.; visualization, S.B., F.L.S. and I.S.; supervision, S.B. and O.S.; project administration, O.S.; funding acquisition, S.B. and O.S. All authors have read and agreed to the published version of the manuscript.

Funding: The project was partially funded by Tallinn University of Technology Development Program 2016–2022, project no. 2014–2020.4.01.16.0032; Tallinn University of Technology, grant no. GFLKSB22; Estonian Research Council, grant no. PRG620. The conference participation was funded by COST Action: CA20101 (PRIORITY).

Institutional Review Board Statement: Not applicable.

Informed Consent Statement: Not applicable.

Data Availability Statement: The data presented in this study are available upon request from the corresponding author.

Acknowledgments: We would like to thank Margit Heinlaan, National Institute of Chemical Physics and Biophysics, Tallinn, for kindly providing us with carboxylated polystyrene microspheres used in this work. We would also like to thank Piotr Garstecki at the Institute of Physical Chemistry, Polish Academy of Sciences, Poland, for the kind provision of surfactant and microfluidic chip mold used in laboratory.

Conflicts of Interest: The authors declare no conflict of interest.

References

1. Veidis, E.M.; LaBeaud, A.D.; Phillips, A.A.; Barry, M. Tackling the Ubiquity of Plastic Waste for Human and Planetary Health. *Am. J. Trop. Med. Hyg.* **2022**, *106*, 12. [CrossRef] [PubMed]
2. Brahney, J.; Mahowald, N.; Prank, M.; Cornwell, G.; Klimont, Z.; Matsui, H.; Prather, K.A. Constraining the atmospheric limb of the plastic cycle. *Proc. Natl. Acad. Sci. USA* **2021**, *118*, e2020719118. [CrossRef] [PubMed]
3. Frias, J.P.G.L.; Nash, R. Microplastics: Finding a consensus on the definition. *Mar. Pollut. Bull.* **2019**, *138*, 145–147. [CrossRef] [PubMed]
4. Gigault, J.; ter Halle, A.; Baudrimont, M.; Pascal, P.-Y.; Gauffre, F.; Phi, T.-L.; El Hadri, H.; Grassl, B.; Reynaud, S. Current opinion: What is a nanoplastic? *Environ. Pollut.* **2018**, *235*, 1030–1034. [CrossRef] [PubMed]
5. Zhang, Y.; Lu, J.; Wu, J.; Wang, J.; Luo, Y. Potential risks of microplastics combined with superbugs: Enrichment of antibiotic resistant bacteria on the surface of microplastics in mariculture system. *Ecotoxicol. Environ. Saf.* **2020**, *187*, 109852. [CrossRef] [PubMed]
6. Imran, M.; Das, K.R.; Naik, M.M. Co-selection of multi-antibiotic resistance in bacterial pathogens in metal and microplastic contaminated environments: An emerging health threat. *Chemosphere* **2019**, *215*, 846–857. [CrossRef] [PubMed]
7. Bartkova, S.; Kahru, A.; Heinlaan, M.; Scheler, O. Techniques Used for Analyzing Microplastics, Antimicrobial Resistance and Microbial Community Composition: A Mini-Review. *Front. Microbiol.* **2021**, *12*, 413. [CrossRef] [PubMed]
8. Berg, S.; Kutra, D.; Kroeger, T.; Straehle, C.N.; Kausler, B.X.; Haubold, C.; Schiegg, M.; Ales, J.; Beier, T.; Rudy, M.; et al. ilastik: Interactive machine learning for (bio)image analysis. *Nat. Methods* **2019**, *16*, 1226–1232. [CrossRef] [PubMed]
9. Carpenter, A.E.; Jones, T.R.; Lamprecht, M.R.; Clarke, C.; Kang, I.H.; Friman, O.; Guertin, D.A.; Chang, J.H.; Lindquist, R.A.; Moffat, J.; et al. CellProfiler: Image analysis software for identifying and quantifying cell phenotypes. *Genome Biol.* **2006**, *7*, R100. [CrossRef] [PubMed]
10. Sanka, I.; Bartkova, S.; Pata, P.; Ernits, M.; Meinberg, M.M.; Agu, N.; Aruoja, V.; Smolander, O.-P.; Scheler, O. User-friendly analysis of droplet array images. *Anal. Chim. Acta* **2023**, *1272*, 341–397. [CrossRef] [PubMed]

proceedings

Abstract

Identification of Emerging Contaminants in the Estonian Aquatic Environment †

Mailis Laht *, Krista Mõts and Laine Parts

Eesti Keskkonnauuringute Keskus OÜ Marja 4d, 10617 Tallinn, Estonia; krista.mots@klab.ee (K.M.); laine.parts@klab.ee (L.P.)
* Correspondence: mailis.laht@klab.ee; Tel.: +372-5063019
† Presented at the International Conference EcoBalt 2023 "Chemicals & Environment", Tallinn, Estonia, 9–11 October 2023.

Keywords: environmental monitoring; target screening; non-target screening

Internationally agreed long-term monitoring indicators do not consider the risks of new synthetic substances. Screening a wide range of chemical substances gives early identification data about substances that exceed ecotoxicological effect values in the aquatic environment. This work summarizes the pollutants found in the Estonian water environment, which can cause both short-term and long-term effects on aquatic life. Data have been collected within the period of 2016–2023 in the framework of various projects, including Estonian national environmental monitoring [1]. Information on substances is collected both with substance-group-based multi-methods (ca 300 substances using HPLC/MS; GC/MS), with wide-scope target screening (2500 substances using LC-ESI-HRMS and GC-APCI-HRMS), and with non-target suspect screening (more than 65,000 compounds in each of the samples including their semi-quantification using LC-ESI-HRMS) [2]. In the screening results, both completely new substances that were not previously associated with environmental risk and substances that have already been regulated and considered an important risk factor were found. In Estonian waters, 10 substances were found in all sea fish and shellfish samples examined. Five of these substances also exceeded the PNEC value on all samples (5′-Methylthioadenosine; 1-Eicosanol, phosphate, compd. with 2,2′-iminobis[ethanol]; Misoprostol; Butyl acrylate; 1-Propanone; 1-(4-dodecyl phenyl)-2-hydroxy-2-methyl-). PAHs and PFASs still pose an environmental risk. The screening identifies regional peculiarities. Not all substances are spread all over the Baltic Sea. Some have significant effects only in Estonian waters. Screening studies of man-made substances found in the environment with the latest analytical methods and knowledge, taking into account the latest scientific developments, will be necessary in the future in order to prevent long-term environmental problems. The presence of substances in the environment depends on the properties of the substance, and therefore it is important to monitor different matrices (water, sediment, biota).

Citation: Laht, M.; Mõts, K.; Parts, L. Identification of Emerging Contaminants in the Estonian Aquatic Environment. *Proceedings* **2023**, 92, 9. https://doi.org/10.3390/proceedings2023092009

Academic Editors: Monika Mortimer, Anne Kahru, Ivo Leito, Riin Rebane and Villem Aruoja

Published: 21 November 2023

Author Contributions: Conceptualization, M.L., K.M. and L.P.; methodology, K.M. and L.P.; validation, M.L., K.M. and L.P.; formal analysis, K.M. and L.P.; investigation, M.L.; resources, M.L.; data curation, M.L.; writing—original draft preparation, M.L.; writing—review and editing, K.M. and L.P.; project administration, M.L.; funding acquisition, M.L. All authors have read and agreed to the published version of the manuscript.

Funding: This research was co-financed by NEFCO BSAP-2021-149, (the Nordic Environment Finance Corporation Baltic Sea Action Plan Fund) and the HELCOM Contracting Parties and Estonian Environmental Ministry.

Institutional Review Board Statement: Not applicable.

Informed Consent Statement: Not applicable.

Data Availability Statement: All data are available in Estonian environmental database "Keskkonnaseire infosüsttem KESE" www.kese.envir.ee.

Acknowledgments: We thank the HELCOM secretariat for their co-operation. For performing the analysis and initial data interpretation, we thank the Environmental Institute (EI), Okruzna 784/42, 97241 Kos, Slovakia, and the National and Kapodistrian University of Athens, Department of Chemistry, Laboratory of Analytical Chemistry, Panepistimiopolis, Athens 15771, Greece.

Conflicts of Interest: The authors declare no conflict of interest.

References

1. Riiklik Keskkonnaseire Programm. Available online: https://kese.envir.ee/kese/welcome.action (accessed on 20 November 2023).
2. HELCOM PreEMP Final Report (Will Be Published by HELCOM Secretariat). Available online: https://helcom.fi/helcom-at-work/projects/pre-empting-pollution-by-screening-for-possible-risks-preempt/) (accessed on 20 November 2023).

proceedings

Abstract

Detection of Mycotoxins and Pyrrolizidine Alkaloids in a Wide Variety of Nutritional Supplements Using the Multianalyte HPLC-MS/MS Method †

Zane Berzina [1,2,*] and Vadims Bartkevics [1,2]

1 Faculty of Chemistry, University of Latvia, Jelgavas iela 1, LV-1004 Riga, Latvia; zane.berzina@lu.lv
2 Institute of Food Safety, Animal Health and Environment "BIOR", Lejupes 3, LV-1076 Riga, Latvia
* Correspondence: zane.berzina@bior.lv

† Presented at the International Conference EcoBalt 2023 "Chemicals & Environment", Tallinn, Estonia, 9–11 October 2023.

Keywords: dietary supplements; mycotoxins; pyrrolizidine alkaloids

Plant-based nutritional supplements are considered a natural means of enriching everyday diets, but they are also a source of various food contaminants, such as mycotoxins [1] and pyrrolizidine alkaloids [2]. Both groups of food contaminants include chemical compounds that can significantly affect consumer health. In this context, the present study aimed to perform a multi-mycotoxin and pyrrolizidine alkaloid (PA) analysis (79 compounds in total) of 47 herbal dietary supplements containing at least one herbal ingredient. The extraction was performed using the QuEChERS method supplemented with the extract freezing-out procedure. Alkaloids were separated using a Luna Omega C18 column and quantified via TSQ Quantiva. The method's LOQs ranged from 0.25 µg kg^{-1} to 500 µg kg^{-1}, and the recoveries ranged from 86% to 119%. The majority of samples contained detectable mycotoxins and PA. Total concentrations ranged up to 5 mg kg^{-1}. High concentrations of alternariol monomethyl ether (AME) and tentoxin were found, with their levels reaching up to 2479 µg kg^{-1} and 307 µg kg^{-1}, respectively. As reported before, many emerging mycotoxins were detected, such as enniatin group mycotoxins and beauvericin, as well as regulated mycotoxins, namely deoxynivalenol, T-2, and HT-2 toxins. Regarding PA, echinatine was determined at the highest concentrations reaching up to 790 µg kg^{-1} (on average 191 µg kg^{-1}), but the total PA concentration in positive samples was in the range of 0.62–1097 µg kg^{-1}. Two samples exceeded the maximum level of 400 µg kg^{-1} for such food supplements. The daily intake of mycotoxins and pyrrolizidine alkaloids may significantly increase with the regular use of such nutritional supplements.

Citation: Berzina, Z.; Bartkevics, V. Detection of Mycotoxins and Pyrrolizidine Alkaloids in a Wide Variety of Nutritional Supplements Using the Multianalyte HPLC-MS/MS Method. *Proceedings* **2023**, *92*, 77. https://doi.org/10.3390/proceedings2023092077

Academic Editors: Monika Mortimer, Anne Kahru, Ivo Leito, Riin Rebane and Villem Aruoja

Published: 14 December 2023

Author Contributions: Z.B.: Conceptualization, methodology, software, validation, formal analysis, investigation, resources, data curation, writing—original draft preparation, visualization; V.B.: supervision, project administration, funding acquisition, writing—review and editing. All authors have read and agreed to the published version of the manuscript.

Funding: This research was funded by Strengthening the doctoral capacity of the University of Latvia within the framework of the new doctoral model, grant number 8.2.2.0/20/1/006.

Institutional Review Board Statement: Not applicable.

Informed Consent Statement: Not applicable.

Data Availability Statement: The data that support the findings of this study are available from the corresponding author, Zane Berzina, upon reasonable request.

Conflicts of Interest: The authors declare no conflict of interest.

References

1. Veprikova, Z.; Zachariasova, M.; Dzuman, Z.; Zachariasova, A.; Fenclova, M.; Slavikova, P.; Hajslova, J. Mycotoxins in Plant-Based Dietary Supplements: Hidden Health Risk for Consumers. *J. Agric. Food Chem.* **2015**, *63*, 6633–6643. [CrossRef] [PubMed]
2. Jansons, M.; Fedorenko, D.; Pavlenko, R.; Berzina, Z.; Bartkevics, V. Nanoflow liquid chromatography mass spectrometry method for quantitative analysis and target ion screening of pyrrolizidine alkaloids in honey, tea, herbal tinctures, and milk. *J. Chrom. A* **2022**, *1676*, 463269. [CrossRef] [PubMed]

proceedings

Abstract

Current Approaches to the Derivatization of Chemical Weapon Convention-Related Alcohol for On-Site Gas Chromatographic Analysis †

Tomas Rozsypal

Nuclear, Biological and Chemical Defence Institute, University of Defence, Vita Nejedleho 1, 68203 Vyskov, Czech Republic; tomas.rozsypal@unob.cz; Tel.: +420-973-452-338

† Presented at the International Conference EcoBalt 2023 "Chemicals & Environment", Tallinn, Estonia, 9–11 October 2023.

Keywords: military identification; chemical warfare agents; silylation; field analysis; environmental sample; urban sample

The task of deployable military laboratories is to perform the fast identification of chemical warfare agents (CWAs) and related chemicals in various types of samples in field conditions. Identification is limited by the time and equipment of mobile laboratories. Polar degradation products are commonly derivatized by a procedure using *N,O*-Bis(trimethylsilyl)trifluoroacetamide (BSTFA) [1,2], which is time-consuming, and the resulting chromatograms often contain a number of artifacts that hinder their identification [3]. This work describes the development and optimization of an alternative trimethylsilylation (TMS) procedure for on-site identification. The goal was to develop a fast and robust method that was efficient without heating, producing clean chromatograms. The analytes were precursors and degradation products of blistering (thiodiglycol, ethyldiethanolamine, methyldiethanolamine, triethanolamine), nerve (*N,N*-diisopropylamino ethanol), and psychoactive (3-quinuclidinol) CWAs. Ten TMS reagents were compared in terms of their derivatization efficiency. The solvent effect, catalyst addition effect, and the time and temperature of derivatization were studied and optimized. The stability of the derivatives was observed over time, and chromatogram artifacts were monitored. The original recommended and widely used method of derivatizing alcohol for 30 min at 60 °C in acetonitrile using BSTFA was overcome using three optimized procedures with different TMS reagents. They achieved high and stable yields in an acetone environment already at room temperature and a reaction lasting 5 min. Due to the same chemical structure of the resulting compounds (TMS-derivatives), it was possible to use established mass spectral databases. Optimized procedures were applied to environmental (water, sand) and urban (acrylic paint, asphalt-aluminum paint, concrete) samples contaminated with the studied alcohol. The results obtained on a benchtop gas chromatograph were validated afterward on a field device used by the deployable chemical laboratory of the Czech Army following their standard operating procedures for sample preparation. The developed methods are useful for military teams and stationary analytical laboratories whose task is the unambiguous identification of CWAs and related compounds in various samples.

Citation: Rozsypal, T. Current Approaches to the Derivatization of Chemical Weapon Convention-Related Alcohol for On-Site Gas Chromatographic Analysis. *Proceedings* **2023**, *92*, 13. https://doi.org/10.3390/proceedings2023092013

Academic Editors: Monika Mortimer, Anne Kahru, Ivo Leito, Riin Rebane and Villem Aruoja

Published: 21 November 2023

Funding: This research received no external funding.

Institutional Review Board Statement: Not applicable.

Informed Consent Statement: Not applicable.

Data Availability Statement: The data presented in this study are available on request from the corresponding author. The data are not publicly available due to internal military regulations.

Acknowledgments: I would like to express my gratitude to the teams of deployable chemical laboratories of the 31st Chemical, Biological, Radiological and Nuclear Regiment of the Czech Army for the possibility of using their capacities to verify the developed methods in the mobile laboratory as part of their regular internship at our Institute under my leadership.

Conflicts of Interest: The author declares no conflict of interest.

References

1. Black, R.; Read, R.; Riches, J. Derivatisation for gas chromatography. In *Recommended Operating Procedures for Analysis in the Verification of Chemical Disarmament*, 2nd ed.; Vanninen, P., Ed.; University of Helsinki: Helsinki, Finland, 2017; pp. 131–152.
2. Witkiewicz, Z.; Sliwka, E.; Neffe, S. Chromatographic analysis of chemical compounds related to the Chemical Weapons Convention. *Trends Anal. Chem.* **2016**, *85*, 21–33. [CrossRef]
3. Rozsypal, T. Development of a method for the derivatization of ethanolamines and its application to sand samples. *J. Serb. Chem. Soc.* **2022**, *87*, 233–245. [CrossRef]

proceedings

MDPI

Abstract

Cost-Effective and Compact Measurement of Arsenic in Water (ARMINE Project-MITY) †

Adriana Ferancova *, Maarit K. Hattuniemi, Veijo Sutinen, Sami Häkkilä, Jay Pee Oña * and Jarkko P. Räty

Measurement Technology Unit, Kajaani University Consortium, University of Oulu, Kehräämöntie 7, 87400 Kajaani, Finland; maarit.hattuniemi@oulu.fi (M.K.H.); veijo.sutinen@oulu.fi (V.S.); sami.hakkila@oulu.fi (S.H.); jarkko.raty@oulu.fi (J.P.R.)

* Correspondence: adriana.ferancova@oulu.fi (A.F.); jay.oa@oulu.fi (J.P.O.)

† Presented at the International Conference EcoBalt 2023 "Chemicals & Environment", Tallinn, Estonia, 9–11 October 2023.

Keywords: arsenic; electrochemistry; water analysis; stripping voltammetry

The contamination of groundwater by arsenic is a serious concern due to the acute and long-term effects of arsenic toxicity on human health. Arsenic contamination is observed in at least 70 countries where the concentration of arsenic in drinking water exceeds the WHO provisional limit of 10 μg/L [1]. In Finland, areas with high concentrations of arsenic in groundwater [2], which affects the quality of well water used for gardening, can be found. Arsenic can enter the food chain when vegetables and fruits are watered with contaminated well water. Although the contamination in Finland mostly originates from natural arsenic, the growing number of mining activities presents a risk of additional pollution from artificial sources. Therefore, proper monitoring of the water supply is necessary to ensure safe levels of arsenic in water for human consumption. Currently, reliable methods for arsenic content determination (e.g., AAS and ICP-MS) are time-consuming and must be carried out under laboratory conditions [3,4]. The aim of our project was to develop a compact and inexpensive method for measuring arsenic concentrations in water samples using an electrochemical sensor. The method involves voltametric stripping [5], which allows for the rapid measurement of arsenic at very low concentrations (ppb levels). A handheld potentiostat with measurement and evaluation software and a mobile phone application were also made. This technology is designed for on-field monitoring of arsenic in industrial and residential areas. The project started with the laboratory-scale development of an electrochemical method for arsenic analysis, which was later implemented at a pilot scale in southern Finland. This study was performed under the ARMINE project of the Measurement Technology (MITY-Kajaani) unit of the University of Oulu (Finland) and is one of the application areas of research for health and clean technology.

Citation: Ferancova, A.; Hattuniemi, M.K.; Sutinen, V.; Häkkilä, S.; Oña, J.P.; Räty, J.P. Cost-Effective and Compact Measurement of Arsenic in Water (ARMINE Project-MITY). *Proceedings* **2023**, *92*, 62. https://doi.org/10.3390/proceedings2023092062

Academic Editors: Monika Mortimer, Anne Kahru, Ivo Leito, Riin Rebane and Villem Aruoja

Published: 29 November 2023

Author Contributions: Conceptualization, A.F., M.K.H. and J.P.O.; methodology, A.F., M.K.H. and J.P.O.; software, S.H. and V.S.; validation, A.F., M.K.H. and J.P.O.; formal analysis, A.F., M.K.H. and J.P.O.; investigation, A.F., M.K.H. and J.P.O.; resources, J.P.R.; data curation, A.F., M.K.H. and J.P.O.; writing—original draft preparation, J.P.O.; writing—review and editing, A.F. and J.P.R.; visualization, A.F., M.K.H. and J.P.O.; supervision, A.F. and J.P.R.; project administration, J.P.R.; funding acquisition, J.P.R. All authors have read and agreed to the published version of the manuscript.

Funding: This research was funded by European Regional Development Fund through North Ostrobothnia Centre for Economic Development, Transport and the Environment (Pohjois-Pohjanmaa ELY) under REACT-EU program, grant number A77835.

Institutional Review Board Statement: Not applicable.

Informed Consent Statement: Not applicable.

Data Availability Statement: The data presented in this study are available upon request from the corresponding author.

Acknowledgments: The authors thank the Pirkkala municipal office, Environmental department, for their recommendations regarding sampling and field test sites. We are also grateful to Harri Vuorenpää for the permission to perform the field test on his land.

Conflicts of Interest: The authors declare no conflict of interest.

References

1. *Guidelines for Drinking-Water Quality*, 4th ed.; World Health Organization: Geneva, Switzerland, 2022; pp. 340–343.
2. Ruskeeniemi, T.; Backman, B.; Loukola-Ruskeeniemi, K.; Sorvari, J.; Lehtinen, H.; Schultz, E.; Mäkelä-Kurtto, R.; Rossi, E.; Vaajasaari, K.; Bilaletdin, Ä. Arsenic in the Pirkanmaa region, southern Finland: From identification through to risk assessment to risk management. *Geol. Surv. Finl. Spec. Pap.* **2011**, *49*, 213–227.
3. Feng, Y.L.; Chen, H.Y.; Tian, L.C.; Narasaki, H. Off-line separation and determination of inorganic arsenic species in natural water by high resolution inductively coupled plasma mass spectrometry with hydride generation combined with reaction of arsenic(V) and L-cysteine. *Anal. Chim. Acta* **1998**, *375*, 167–175. [CrossRef]
4. Anezaki, K.; Nukatsuka, I.; Ohzeki, K. Determination of Arsenic(III) and Total Arsenic(III,V) in Water Samples by Resin Suspension Graphite Furnace Atomic Absorption Spectrometry. *Anal. Sci.* **1999**, *15*, 829–834. [CrossRef]
5. Mays, D.E.; Hussam, A. Voltammetric methods for determination and speciation of inorganic arsenic in the environment—A review. *Anal. Chim. Acta* **2009**, *646*, 6–16. [CrossRef] [PubMed]

proceedings

Abstract

From Organometallic Chemistry to Multifunctional Nanoparticle-Based Devices for Gas Detection and Degradation of Air Pollutants †

Martin Jakoobi [1], Katia Fajerweg [1,2] and Myrtil L. Kahn [1,*]

1 Laboratory of Coordination Chemistry (LCC), CNRS UPR 8241, University of Toulouse, 205 Route de Narbonne, 31077 Toulouse, France; martin.jakoobi@lcc-toulouse.fr (M.J.); katia.fajerweg@lcc-toulouse.fr (K.F.)
2 Université de Toulouse, UT3 Paul Sabatier, 118 Route de Narbonne, CEDEX 04, 31062 Toulouse, France
* Correspondence: myrtil.kahn@lcc-toulouse.fr
† Presented at the International Conference EcoBalt 2023 "Chemicals & Environment", Tallinn, Estonia, 9–11 October 2023.

Keywords: organometallic chemistry; nanoparticles; mild conditions; size control; pollutants degradation

Considering the climate, societal and health-related current and emerging issues facing the world, our group, as part of the (nano-)material science community, will play a part in providing materials and technology that can tackle these issues. Our strategy focuses on the design and development of complex hybrid nano-objects and nanomaterials with unprecedented properties, with the aim of developing functional and innovative solutions to societal challenges (Figure 1). To achieve this, we are applying an organometallic approach for the synthesis of well-defined nanoparticles (NPs) and nanomaterials [1]. This bottom-up approach allows control of the NPs synthesis (size, shape, colloidal stability) on a molecular level with the help of cleverly designed starting molecular precursor(s), under mild reaction conditions and in safe-by-design approaches [2]. The presentation will focus on our team's research related to the synthesis and properties of NPs and nanomaterials, their implementation into devices for either gas detection (i.e., sensors based on Cu [3], Zn [4,5], and Sn oxide NPs [6,7]) or degradation of air pollutants [8], and the interconnection between different fields (chemistry, physical chemistry, physics, and biology).

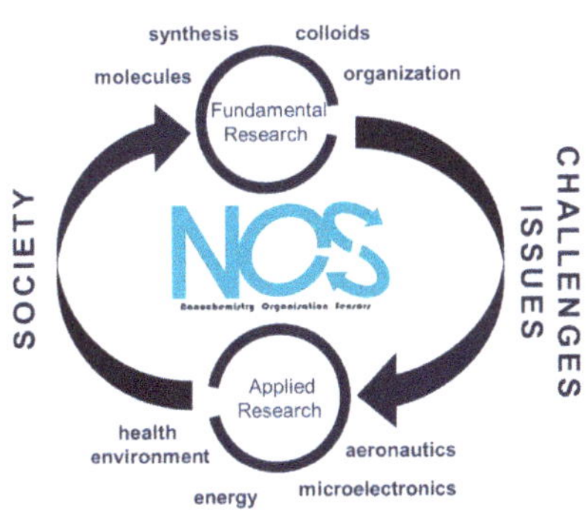

Figure 1. Positive feedback loop on the development of new nanomaterials and technological solutions to societal needs.

Citation: Jakoobi, M.; Fajerweg, K.; Kahn, M.L. From Organometallic Chemistry to Multifunctional Nanoparticle-Based Devices for Gas Detection and Degradation of Air Pollutants. *Proceedings* **2023**, 92, 42. https://doi.org/10.3390/proceedings2023092042

Academic Editors: Monika Mortimer, Anne Kahru, Ivo Leito, Riin Rebane and Villem Aruoja

Published: 24 November 2023

Author Contributions: Writing—original draft preparation, M.J.; writing—review and editing, K.F. and M.L.K.; supervision, K.F. and M.L.K. All authors have read and agreed to the published version of the manuscript.

Funding: This research was funded by the Centre National de la Recherche Scientifique, CNRS and Université de Toulouse.

Institutional Review Board Statement: Not applicable.

Informed Consent Statement: Not applicable.

Data Availability Statement: The data presented during the oral presentation was based on the references above, from where relevant research data can be found.

Conflicts of Interest: The authors declare no conflict of interest.

References

1. Amiens, C.; Chaudret, B.; Ciuculescu-Pradines, D.; Collière, V.; Fajerwerg, K.; Fau, P.; Kahn, M.; Maisonnat, A.; Soulantica, K.; Philippot, K. Organometallic approach for the synthesis of nanostructures. *New J. Chem.* **2013**, *37*, 3374–3401. [CrossRef]
2. Carnide, C.; Champouret, Y.; Valappil, D.; Vahlas, C.; Mingotaud, A.-F.; Clergereaux, R.; Kahn, M.L. Secured Nanosynthesis–Deposition Aerosol Process for Composite Thin Films Incorporating Highly Dispersed Nanoparticles. *Adv. Sci.* **2022**, *10*, 2204929. [CrossRef] [PubMed]
3. Jońca, J.; Ryzhikov, A.; Palussière, S.; Esvam, J.; Fajerwerg, K.; Menini, P.; Kahn, M.L.; Fau, P. Organometallic Synthesis of CuO Nanoparticles: Application in Low-Temperature CO Detection. *ChemPhysChem* **2017**, *18*, 2658–2665. [CrossRef] [PubMed]
4. Jońca, J.; Ryzhikov, A.; Kahn, M.L.; Fajerwerg, K.; Chaudret, B.; Chappelle, A.; Menini, P.; Fau, P. Shape-controlled ZnO Nanistructures for Gas Sensing Applications. *Procedia Eng.* **2014**, *87*, 907–910. [CrossRef]
5. Ryzhikov, A.; Jońca, J.; Kahn, M.; Fajerwerg, K.; Chaudret, B.; Chappelle, A.; Menini, P.; Shim, C.H.; Gaudon, A.; Fau, P. Organometallic synthesis of ZnO nanoparticles for gas sensing: Towards selectivity through nanoparticles morphology. *J. Nanopart. Res.* **2015**, *17*, 280. [CrossRef]
6. Jońca, J.; Ryzhikov, A.; Fajerwerg, K.; Kahn, M.L.; Chaudret, B.; Chappelle, A.; Menini, P.; Fau, P. A Novel SnO_2 Sensor and its Selectivity Improvement with Catalytic Filter. *Procedia Eng.* **2014**, *87*, 923–926. [CrossRef]
7. Jońca, J.; Ryzhikov, A.; Kahn, M.L.; Fajerwerg, K.; Chappelle, A.; Menini, P.; Fau, P. SnO_2 "Russian Doll" Octahedra Prepared by Metalorganic Synthesis: A New Structure for Sub-ppm CO Detection. *Chem. Eur. J.* **2016**, *22*, 10127–10135. [CrossRef] [PubMed]
8. Castelló Lux, K.; Hot, J.; Fau, P.; Bertron, A.; Kahn, M.L.; Ringot, E.; Fajerwerg, K. Nano-gold decorated ZnO: An alternative photocatalyst promising for NO_x degradation. *Chem. Eng. Sci.* **2023**, *267*, 118377. [CrossRef]

proceedings

Abstract

Development of a Harmonised Methodology for Narva River Water Discharge Estimation †

Yaroslav Kobets * and Alvina Reihan

Department of Civil Engineering and Architecture, Tallinn University of Technology, Ehitajate tee 5, 19086 Tallinn, Estonia; alvina.reihan@taltech.ee
* Correspondence: yaroslav.kobets@taltech.ee

† Presented at the International Conference EcoBalt 2023 "Chemicals & Environment", Tallinn, Estonia, 9–11 October 2023.

Keywords: river discharge estimation; transboundary rivers; Narva River; rating curves

The Narva River has its source in Lake Peipsi and flows into the Gulf of Finland. It passes through Estonia and Russia along its course. Estonia is a member of the European Union while Russia is not. In addition to this, the two countries use completely different techniques for discharge estimation. This all creates difficulties in river runoff calculations, affecting calculations of pollution load, providing data inconsistencies when the countries report to the Helsinki Commission (HELCOM) as they are required to. Besides this, the river has a unique feature called the "backwater effect", which affects the stage–discharge relationships; thus, the usage of simple stage–discharge rating curves leads to a lot of errors.

The differences between the discharge numbers from two countries can reach up 80% for certain months.

The NARVAWATMAN project, which was implemented under the European Neighbourhood Instrument and co-financed by the European Union, strived to analyse the existing discharge estimation techniques, conduct additional field measurements, come up with a new harmonised solution, and provide recommendations for future analysis in the runoff estimation of the Narva River [1].

Citation: Kobets, Y.; Reihan, A. Development of a Harmonised Methodology for Narva River Water Discharge Estimation. *Proceedings* **2023**, *92*, 18. https://doi.org/10.3390/proceedings2023092018

Academic Editors: Monika Mortimer, Anne Kahru, Ivo Leito, Riin Rebane and Villem Aruoja

Published: 21 November 2023

Author Contributions: Conceptualization and methodology, Y.K. and A.R.; formal analysis, Y.K. and A.R.; data curation, Y.K.; writing—original draft preparation Y.K.; project administration, A.R. All authors have read and agreed to the published version of the manuscript.

Funding: This research was funded by the Project ER25 NarvaWatMan (Water Management of the Narva River: harmonization and sustention) under the European Neighbourhood Instrument and co-financed by the European Union (https://estoniarussia.eu/ (accessed on 1st of September 2023)). The Taltech ID of this project is V19016.

Institutional Review Board Statement: Not applicable.

Informed Consent Statement: Not applicable.

Data Availability Statement: The data presented in this study are available on request from the corresponding author. The data are not publicly available due to its being a part of the project activity and it wasn't planned to grant public access.

Acknowledgments: This study was carried out within the project "Water management of the Narva River: harmonization and sustention", funded by the Estonian EU external border programme for 2014–2020 and supported by the partners.

Conflicts of Interest: The authors declare no conflict of interest. The funders had no role in the design of the study; in the collection, analyses, or interpretation of data; in the writing of the manuscript; or in the decision to publish the results.

Reference

1. Water Environmental Engineering Research Group. *NarvaWatMan Project, Final Report*; Water Environmental Engineering Research Group: Tallinn, Estonia, 2022.

proceedings

Abstract

Assessing Biological Effects of Contaminants in the Gulf of Finland, Northeastern Baltic Sea, Using Sediment Biotests with Amphipods (*Monoporeia affinis*) and Biomarker Responses in Clams (*Macoma balthica*) †

Ivan Kuprijanov [1,*], Natalja Kolesova [1], Maarja Lipp [1] and Kari K. Lehtonen [2]

1 Department of Marine Systems, Tallinn University of Technology, 19086 Tallinn, Estonia; natalja.kolesova@taltech.ee (N.K.); maarja.lipp@taltech.ee (M.L.)
2 Marine and Freshwater Solutions Unit, Finnish Environment Institute, 00790 Helsinki, Finland; kari.lehtonen@syke.fi
* Correspondence: ivan.kuprijanov@taltech.ee; Tel.: +372-6204342

† Presented at the International Conference EcoBalt 2023 "Chemicals & Environment", Tallinn, Estonia, 9–11 October 2023.

Keywords: Gulf of Finland; biological effects; chemical contamination; sediment biotests; biomarkers; *Monoporeia affinis*; *Macoma balthica*

The Gulf of Finland, in the northeastern Baltic Sea, is experiencing ongoing adverse effects due to human activities, leading to a decline in the quality of the marine environment [1]. The current emphasis in environmental monitoring and assessment lies in chemical and ecological measurements, with little attention given to the connection between these measurements and their biological effects. The neglect of examining biological effects hampers our understanding of the overall influence that various contaminants have on marine organisms, which results from complex combinations of multiple effects. We have collected sediments from moderately to highly contaminated offshore and coastal areas with subsequent analyses of selected chemicals (Figure 1). Where available, clams (*M. balthica*) were collected for biological effects measurements. From seven sites, whole-sediment bioassays with amphipods (*M. affinis*) were conducted to determine the effect of contaminants with the registration of the mortality rate and activity of three biochemical biomarkers. In the sediment biotest, the mortality rate was mostly uniformly low (around 8%). The comparison of the amphipod and clam biomarker data revealed that the amphipod, which was exposed to sediments from Narva Bay, did not exhibit significant changes in biomarker activities, except for catalase (CAT), which indicates oxidative stress (Figure 2). In clams, peaks and falls in enzymatic activities primarily reflect in situ exposure to harmful compounds and conditions. The lowest glutathione S-transferase (GST) activity in clams might be related to the impact of contaminants, as high levels of mercury registered simultaneously in the sediments near the Narva river mouth, while near Kunda harbour, the normalised content of PAH anthracene exceeded more than five times the HELCOM threshold. The highest GST in Narva bay clams might be related to the mixed impact of toxic biocide TBT, which exceeded the GES threshold by almost ten times, and moderate contamination by PAHs and non-dioxin-like PCBs was found in the sediments there. According to the calculated integrated biomarker response index, the highest value at the Sillamäe harbour reflects the most stressful conditions within the studied area. In addition, the elevated level of oxidative stress hints at the unfavourable hydrophysical and chemical conditions in this location.

Citation: Kuprijanov, I.; Kolesova, N.; Lipp, M.; Lehtonen, K.K. Assessing Biological Effects of Contaminants in the Gulf of Finland, Northeastern Baltic Sea, Using Sediment Biotests with Amphipods (*Monoporeia affinis*) and Biomarker Responses in Clams (*Macoma balthica*). *Proceedings* **2023**, 92, 54. https://doi.org/10.3390/proceedings2023092054

Academic Editors: Monika Mortimer, Anne Kahru, Ivo Leito, Riin Rebane and Villem Aruoja

Published: 27 November 2023

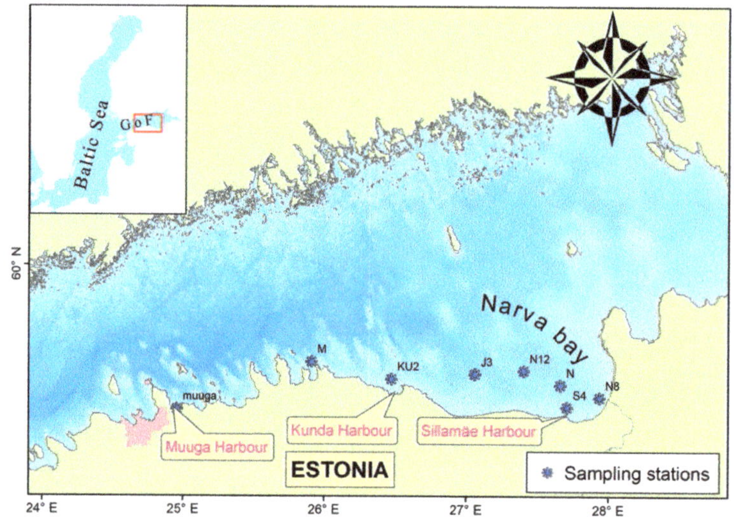

Figure 1. Sampling stations in the study area.

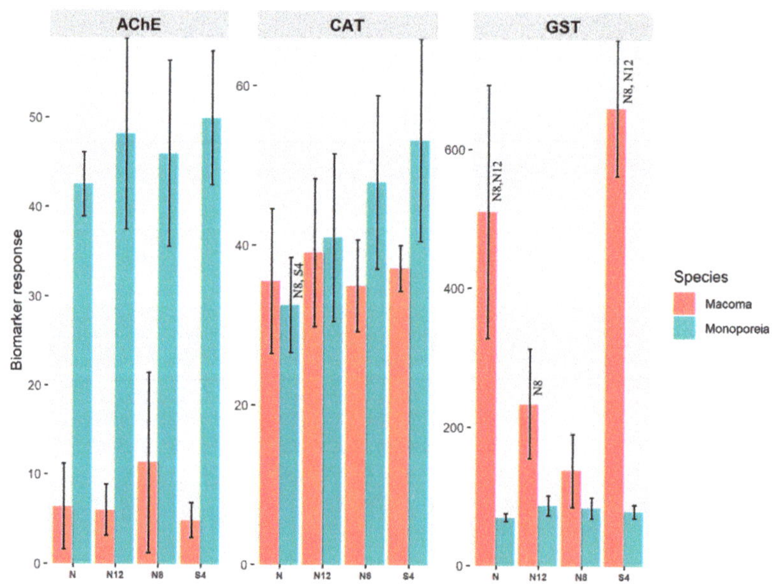

Figure 2. Biomarker response in both species (mean ± sd). Station names above the bars indicate significant differences between stations ($p < 0.05$).

Author Contributions: Conceptualization, I.K.; methodology, I.K., N.K., M.L. and K.K.L.; formal analysis, I.K.; investigation, I.K., N.K. and M.L.; writing—original draft preparation, I.K.; writing—review and editing, I.K. and K.K.L.; supervision, I.K. and K.K.L.; project administration, I.K.; funding acquisition, I.K., N.K. and K.K.L. All authors have read and agreed to the published version of the manuscript.

Funding: This research was funded by the European Neighbourhood Instrument and co-financed by the European Union (HAZLESS, NarBaltAware) (grant numbers: ER90/194), by the European Biodiversity Partnership Biodiversa + (D2P) (grant number: 2021-473), and the Environmental Investment Centre (KIK 17253).

Institutional Review Board Statement: Not applicable.

Informed Consent Statement: Not applicable.

Data Availability Statement: Available on request.

Acknowledgments: We want to thank our colleagues from the Department of Marine Systems (TalTech), namely, Urmas Lips, Fred Buschmann, Oliver Sammlas, and the crew of the R/V Salme, for their comprehensive support in the organization and conduction of the fieldworks.

Conflicts of Interest: The authors declare no conflict of interest. The funders had no role in the design of the study; in the collection, analyses, or interpretation of data; in the writing of the manuscript; or in the decision to publish the results. The content of this publication is the sole responsibility of authors and can under no circumstances be regarded as reflecting the position of the Programme-participating countries or the European Union.

Reference

1. Kuprijanov, I.; Väli, G.; Sharov, A.; Berezina, N.; Liblik, T.; Lips, U.; Kolesova, N.; Maanio, J.; Junttila, V.; Lips, I. Hazardous substances in the sediments and their pathways from potential sources in the eastern Gulf of Finland. *Mar. Pollut. Bull.* **2021**, *170*, 112642. [CrossRef]

proceedings

Abstract

Estimation of the Share of Total Nutrient Load from the Territory of Estonia along the Narva River to the Baltic Sea †

Alvina Reihan [1,*] and Kati Roosalu [2]

1 Department of Civil Engineering and Architecture, Tallinn University of Technology, Ehitajate tee 5, 19086 Tallinn, Estonia
2 Department of Energy Technology, Tallinn University of Technology, Ehitajate tee 5, 19086 Tallinn, Estonia; kati.roosalu@taltech.ee
* Correspondence: alvina.reihan@taltech.ee
† Presented at the International Conference EcoBalt 2023 "Chemicals & Environment", Tallinn, Estonia, 9–11 October 2023.

Keywords: nitrogen and phosphorus; nutrient load estimation; transboundary rivers; Narva River; Baltic Sea

The annual inputs of nitrogen and phosphorus by the Narva river into the Baltic Sea are distributed between Estonia and Russia according to the catchment area shares of 33% and 67%, respectively [1]. A comparison of the total nutrient load shows that the proportion between countries may vary depending on what input data and estimation methods are used. The aim of this study was to improve the comparability of estimates in order to obtain more accurate information on the distribution of nutrients from Estonia. The results were obtained from the testing of the Estonian and Russian calculated methods [2] and initial data for the period 2006–2018, as well as the distribution of sources calculated for the direct catchment area of the transboundary Narva River. Detailed calculations for each sub-basin were carried out using mainly national coefficients. A comparison of the calculated total input with the monitoring data showed that both countries' estimates of nitrogen and phosphorus loads correlated quite well with the monitoring data. The main sources of Ntot in the immediate catchment were point sources and agriculture. The main difference was in the Ptot input from natural and urban areas and agriculture. Therefore, point sources, agriculture and runoff from urban areas can be considered the priority sources. According to the results obtained for the Estonian territory, the contribution of Ntot was 23% and that of Ptot was 29% [3].

Citation: Reihan, A.; Roosalu, K. Estimation of the Share of Total Nutrient Load from the Territory of Estonia along the Narva River to the Baltic Sea. *Proceedings* **2023**, *92*, 44. https://doi.org/10.3390/proceedings2023092044

Academic Editors: Monika Mortimer, Anne Kahru, Ivo Leito, Riin Rebane and Villem Aruoja

Published: 24 November 2023

Author Contributions: Conceptualization and methodology, A.R.; formal analysis, A.R. and K.R.; data curation, A.R. and K.R.; writing—original draft preparation, A.R.; project administration, A.R. All authors have read and agreed to the published version of the manuscript.

Funding: This research (Project Water Management of the Narva River: Harmonization and Sustention) is implemented under the European Neighbourhood Instrument and co-financed by the European Union (https://estoniarussia.eu/ (accessed on 1 September 2023)). This study was conducted at Taltech with project IDV19016.

Institutional Review Board Statement: Not applicable.

Informed Consent Statement: Not applicable.

Data Availability Statement: The data presented in this study are available on request from the corresponding author. The data are not publicly available due to its being a part of the project activity and it wasn't planned to grant public access.

Acknowledgments: The authors acknowledge the assistance of the Estonian EU external border programme for 2014–2020 and it's the project partners for their contribution.

Conflicts of Interest: The authors declare no conflict of interest. The funders had no role in the design of the study; in the collection, analyses, or interpretation of data; in the writing of the manuscript; or in the decision to publish the results.

References

1. HELCOM, 2018. The Sixth Pollution Load Compilation (PLC-6). Available online: https://helcom.fi/wp-content/uploads/2019/08/PLC-6-Executive-Summary.pdf (accessed on 31 October 2020).
2. Applied Methodology for the PLC-6 Assessment Edited by Lars M. Svendsen, DCE Aarhus University, PLC-6 Project Manager, 60p. Available online: https://helcom.fi/wp-content/uploads/2020/01/PLC-6-methodology.pdf (accessed on 15 December 2021).
3. Water Environmental Engineering Research Group. *Narva Wat Man Project*; Final Report; Taltech: Tallinn, Estonia, 2022.

proceedings

Abstract

Water Quality Index (WQI) as a Tool to Assess Waterbody Status: Joint WQI Model Development for River Narva and the Rivers of the Lake Peipsi Basin (Based on Results of the NarvaWatMan Project) †

Kati Roosalu

Department of Energy Technology, Tallinn University of Technology, Ehitajate tee 5, 19086 Tallinn, Estonia; kati.roosalu@taltech.ee

† Presented at the International Conference EcoBalt 2023 "Chemicals & Environment", Tallinn, Estonia, 9–11 October 2023.

Keywords: WQI; WFD; pollution load; transboundary rivers; Narva River; Baltic Sea

One of the aims of The European Water Framework Directive (WFD) is that the Member States monitor and assess the ecological and chemical status of all surface water bodies (rivers, lakes, transitional and coastal waters) and to ensure that they achieve a good status by 2027, with an interim deadline of 2021. The question is, what is meant by a good status and how to assess it? A good status of a water body can depend on several factors such as its physical, chemical, and biological characteristics, which can mean a lot of data. The water quality index (WQI) is a tool to provide a numerical value that indicates the overall health of the water in a particular area. A summary of how the WQI has developed and how it can be used was provided. When managing joint water bodies, it is especially important that decisions are made on a uniform basis, and the status of the water body is assessed in a similar way. During the project NarvaWatMan, the comparison was made of how the status of the river Narva is assessed both in Estonia and Russia. Based on the collected data, a joint water quality index was developed, and the received results were presented [1].

Citation: Roosalu, K. Water Quality Index (WQI) as a Tool to Assess Waterbody Status: Joint WQI Model Development for River Narva and the Rivers of the Lake Peipsi Basin (Based on Results of the NarvaWatMan Project). *Proceedings* **2023**, *92*, 29. https://doi.org/10.3390/proceedings2023092029

Academic Editors: Monika Mortimer, Anne Kahru, Ivo Leito, Riin Rebane and Villem Aruoja

Published: 22 November 2023

Funding: This research was funded by the Project ER25 NarvaWatMan (Water Management of the Narva River: harmonization and sustention) under the European Neighbourhood Instrument and co-financed by the Estonian EU external border programme for 2014–2020 (https://estoniarussia.eu/ accessed on 21 November 2023), under TalTech project number V19016 and supported by the partners.

Institutional Review Board Statement: Not applicable.

Informed Consent Statement: Not applicable.

Data Availability Statement: Publicly available datasets were analyzed in this study. This data can be found here: [https://kese.envir.ee] (accessed on 21 November 2023).

Conflicts of Interest: The author declares no conflict of interest. The funders had no role in the design of the study; in the collection, analyses, or interpretation of data; in the writing of the manuscript; or in the decision to publish the results.

Reference

1. *NarvaWatMan Project*; Final Report, Water Environmental Engineering Research Group. By special request: Alvina.reihan@taltech.ee; TalTech: Tallinn, Estonia, 2022.

proceedings

Abstract

Non-Formal Ecological Education: Innovative Methods Tested in Lake Peipsi Communities †

Margit Säre

Peipsi Center for Transboundary Cooperation, 51009 Tartu, Estonia; margitsare@gmail.com

† Presented at the International Conference EcoBalt 2023 "Chemicals & Environment", Tallinn, Estonia, 9–11 October 2023.

Keywords: non-formal ecological education; innovative environmental education methods; Lake Peipsi region

Our continent faces serious problems in areas such as biodiversity loss, limited natural resources, the impacts of climate change, etc. Thus, during the last decade, environmental education gained importance globally as the "alarm clock" that is now ringing loud, and educators, as well as politicians, understand the urgent need to better educate young people on the complexity and interconnection of ecological, economic, political, cultural issues and develop creative problem-solving skills. This requires new types of educational methods, as recent research also suggests that non-formal and interactive approaches to ecological education are more effective than purely fact-based teaching. This article describes innovative, interactive methods tested by the Estonian NGO Peipsi Center for Transboundary Cooperation in regional schools and also during public events. Deliverables such as textbooks/worksheets on Peipsi ecosystems, the LoqQuiz orientation game, crossword and educational videos with quizzes on Peipsi ecology, and Educational Live-Action Role Play (EduLARP) were among the tasks that had the highest interest among our target group [1], Our experience suggests that interactive and narrative approaches in environmental education such as problem-solving exercises and role play can be more effective in the long run and engage pupils better than fact-based (classroom) teaching. Non-formal learning is characterized as learning by doing, foremost from the specific situations that pupils experience, while their attitudes and values shape their future behavior [2]. Also, ecological education should focus more on the importance of individual responsibility as well as the role of collective action, and it is very important to constantly work with educators, especially in more peripheral areas.

Citation: Säre, M. Non-Formal Ecological Education: Innovative Methods Tested in Lake Peipsi Communities. *Proceedings* **2023**, *92*, 83. https://doi.org/10.3390/proceedings2023092083

Academic Editors: Monika Mortimer, Anne Kahru, Ivo Leito, Riin Rebane and Villem Aruoja

Published: 24 January 2024

Funding: The preparation of this article was supported by the Estonian EU external border programme project ER194 "The Narva River, from Lake Peipsi to the Baltic Sea: Challenges and Opportunities" and Estonian International Development Centre funded project "Testing methods to develop Global Citizen competences model".

Institutional Review Board Statement: Not applicable.

Informed Consent Statement: Not applicable.

Data Availability Statement: Not applicable.

Conflicts of Interest: The author affiliated with Peipsi Center for Transboundary Cooperation declares no conflict of interest. The funders had no role in the design of the study; in the collection, analyses, or interpretation of data; in the writing of the manuscript; or in the decision to publish the results.

References

1. Formal, Non-Formal and Informal Learning, Council on Europe. Available online: https://www.coe.int/en/web/lang-migrants/formal-non-formal-and-informal-learningL (accessed on 20 June 2023).
2. Publications, Peipsi Center for Transboundary Cooperation. Available online: https://ctc.ee/publications (accessed on 10 June 2023).

proceedings

MDPI

Abstract

Nanocomposite Metal Oxide/Hydroxide Adsorbents for Advanced Wastewater Treatment and Toxicological Risk Assessment for the Aquatic Environment †

Asya Drenkova-Tuhtan [1,*], Mariliis Sihtmäe [1], Irina Blinova [1], Kevin Uke [2], Heiki Vija [1] and Anne Kahru [1,3]

1 National Institute of Chemical Physics and Biophysics, 12618 Tallinn, Estonia; mariliis.sihtmae@kbfi.ee (M.S.); irina.blinova@kbfi.ee (I.B.); heiki.vija@kbfi.ee (H.V.); anne.kahru@kbfi.ee (A.K.)
2 Nordica | Nordic Aviation Group AS, 11101 Tallinn, Estonia; kevinuke95@gmail.com
3 Estonian Academy of Sciences, 10130 Tallinn, Estonia
* Correspondence: asya.drenkova@kbfi.ee

† Presented at the International Conference EcoBalt 2023 "Chemicals & Environment", Tallinn, Estonia, 9–11 October 2023.

Keywords: aquatic toxicity; magnetite; metal-based adsorbents; phosphorus recovery; phosphorus removal; safe-by-design; nanomaterials; wastewater effluent polishing; zinc

Phosphorus (P) is a key nutrient for agriculture [1], but is also an environmental pollutant that causes eutrophication and is commonly removed from wastewater [2]. Engineered nanostructured materials, predominantly metal oxides/hydroxides, are frequently reported as excellent adsorbents for phosphate [3], able to selectively remove P from wastewater to ultra-low concentrations [4], and facilitate P-recovery through reversible sorption [5]; however, their environmental safety is rarely addressed. This study assesses the ecotoxicological hazard of 10 highly efficient metal oxide/hydroxide nanocomposite P-adsorbents using toxicity tests involving two different food-web level test organisms: the naturally bioluminescent marine bacterium *Vibrio fischeri*, and the crustacean *Daphnia magna*. Nanocomposites were synthesized based on published procedures [6,7] via co-precipitation of 2-, 3- and 4-valent metal precursors (Zn^{2+}, Ca^{2+}, Mg^{2+}, Fe^{3+}, Zr^{4+}) at different molar ratios, and characterized with laser diffraction, ICP-OES, XRD and SEM. Among these, the pilot-scale tested ZnFeZr-6:1:1-oxyhydroxide [8] was modified by reducing the zinc fraction to minimize leaching of toxic Zn^{2+} ions. The composites' stability was investigated in deionized water and 2% NaCl (*V. fischeri* test medium), addressing agglomeration, settling and solubilization (the release of metal ions and/or potentially hazardous nanoparticles). All composites, their filtered supernatants and precursor metal salts were evaluated for their toxic potency (half-effective concentration, EC_{50} and minimum bactericidal concentration, MBC) using three different tests: a *Vibrio fischeri* 30 min bioluminescence inhibition assay (ISO-21338:2010) [9], a *V. fischeri* 24 h viability assay ('Spot test') and a *Daphnia magna* 48 h acute immobilization test (OECD-202) [10]. Only the Zn-containing composites showed inhibitory effects on both organisms. Those with the highest zinc fraction (ZnFeZr-18:5:1; ZnFeZr-10:1:1) were classified "harmful" to *V. fischeri* ($10 < EC_{50} \leq 100$ mg/L), and toxic to *D. magna* ($1 < EC_{50} \leq 10$ mg/L); therefore, they are environmentally unsafe for engineering applications. The ZnFeZr-6:1:1 (*V. fischeri* EC_{50} = 118 mg/L; *D. magna* EC_{50} = 7.7 mg/L) proved assumingly safe for both aquatic organisms once deposited on magnetic particles ZnFeZr-6:1:1@MPs (EC_{50} >> 100 mg/L, MBC > 1000 mg/L). All other composites without Zn were non-toxic, both to *V. fischeri*, and to the more sensitive *D. magna*.

Citation: Drenkova-Tuhtan, A.; Sihtmäe, M.; Blinova, I.; Uke, K.; Vija, H.; Kahru, A. Nanocomposite Metal Oxide/Hydroxide Adsorbents for Advanced Wastewater Treatment and Toxicological Risk Assessment for the Aquatic Environment. *Proceedings* **2023**, *92*, 28. https://doi.org/10.3390/proceedings2023092028

Academic Editors: Monika Mortimer, Ivo Leito, Riin Rebane and Villem Aruoja

Published: 22 November 2023

Author Contributions: Conceptualization, A.D.-T. and A.K.; methodology, M.S., I.B. and H.V.; software, M.S., K.U. and H.V.; validation, M.S., I.B. and H.V.; formal analysis, A.D.-T. and K.U.; investigation, A.D.-T. and K.U.; resources, M.S., I.B. and A.K.; data curation, A.D.-T. and K.U.; writing—original draft preparation, A.D.-T.; writing—review and editing, A.K.; visualization, A.D.-T. and K.U.; supervision, A.K.; project administration, A.D.-T. and A.K.; funding acquisition, A.D.-T. All authors have read and agreed to the published version of the manuscript.

Funding: This research was funded by the EUROPEAN COMMISSION, Grant agreement ID: 867457, through an individual fellowship within the Marie Skłodowska-Curie MSCA-IF-EF-ST funding scheme of the H2020-EU.4 program (spreading excellence and widening participation). This work was also supported through grant PRG749 and conducted using the NAMUR+ core facility (TT13), both funded by the Estonian Research Council.

Institutional Review Board Statement: Not applicable.

Informed Consent Statement: Not applicable.

Data Availability Statement: Data sharing is not applicable to this article.

Acknowledgments: We gratefully acknowledge the Particle Technology group at Fraunhofer Institute for Silicate Research (ISC) in Würzburg, Germany for guiding the materials synthesis process, and for the analytical and technical support through sample analyses and interpretation of materials data. We also acknowledge Triin Vaimann for the assistance with the experimental work on *D. magna*.

Conflicts of Interest: The authors declare no conflict of interest. Nordic Aviation Group AS declares no conflict of interest. The funders had no role in the design of the study; in the collection, analyses, or interpretation of data; in the writing of the manuscript; or in the decision to publish the results.

References

1. Withers, P.J.A.; Forber, K.G.; Lyon, C.; Rothwell, S.; Doody, D.G.; Jarvie, H.P.; Martin-Ortega, J.; Jacobs, B.; Cordell, D.; Patton, M.; et al. Towards resolving the phosphorus chaos created by food systems. *Ambio* **2020**, *49*, 1076–1089. [CrossRef]
2. Dodds, W.K.; Bouska, W.W.; Eitzmann, J.L.; Pilger, T.J.; Pitts, K.L.; Riley, A.J.; Schloesser, J.T.; Thornbrugh, D.J. Eutrophication of U.S. freshwaters: Analysis of potential economic damages. *Env. Sci. Technol.* **2009**, *43*, 12–19. [CrossRef] [PubMed]
3. Ooi, K.; Sonoda, A.; Makita, Y.; Torimura, M. Comparative study on phosphate adsorption by inorganic and organic adsorbents from a diluted solution. *J. Env. Chem. Eng.* **2017**, *5*, 3181–3189. [CrossRef]
4. Drenkova-Tuhtan, A.; Sheeleigh, E.K.; Rott, E.; Meyer, C.; Sedlak, D.L. Sorption of recalcitrant phosphonates in reverse osmosis concentrates and wastewater effluents—influence of metal ions. *Water Sci. Technol.* **2021**, *83*, 934–947. [CrossRef]
5. Drenkova-Tuhtan, A.; Schneider, M.; Mandel, K.; Meyer, C.; Gellermann, C.; Sextl, G.; Steinmetz, H. Influence of cation building blocks of metal hydroxide precipitates on their adsorption and desorption capacity for phosphate in wastewater—A screening study. *Colloids Surf. A Physicochem. Eng. Asp.* **2016**, *488*, 145–153. [CrossRef]
6. Mandel, K.; Drenkova-Tuhtan, A.; Hutter, F.; Gellermann, C.; Steinmetz, H.; Sextl, G. Layered double hydroxide ion exchangers on superparamagnetic microparticles for recovery of phosphate from waste water. *J. Mater. Chem. A* **2013**, *1*, 1840–1848. [CrossRef]
7. Schneider, M.; Drenkova-Tuhtan, A.; Szczerba, W.; Gellermann, C.; Meyer, C.; Steinmetz, H.; Mandel, K.; Sextl, G. Nanostructured ZnFeZr oxyhydroxide precipitate as efficient phosphate adsorber in waste water: Understanding the role of different material-building-blocks. *Environ. Sci. Nano* **2017**, *4*, 180–190. [CrossRef]
8. Drenkova-Tuhtan, A.; Schneider, M.; Franzreb, M.; Meyer, C.; Gellermann, C.; Sextl, G.; Mandel, K.; Steinmetz, H. Pilot-scale removal and recovery of dissolved phosphate from secondary wastewater effluents with reusable ZnFeZr adsorbent@Fe_3O_4/SiO_2 particles with magnetic harvesting. *Water Res.* **2017**, *109*, 77–87. [CrossRef] [PubMed]
9. *ISO 21338:2010*; Water Quality—Kinetic Determination of the Inhibitory Effects of Sediment, Other Solids and Coloured Samples on the Light Emission of Vibrio Fischeri. International Organization for Standardization: Geneva, Switzerland, 2010.
10. *OECD 202*; Guideline for Testing of Chemicals: Daphnia sp., Acute Immobilisation Test. OECD: Paris, France, 2004.

proceedings

MDPI

Abstract

Silver–Chitosan Nanocomposites for Biomedical Application: Design, Synthesis and Antimicrobial Efficiency †

Kaja Kasemets [1,*], Jüri Laanoja [1,2], Mariliis Sihtmäe [1], Heiki Vija [1], Imbi Kurvet [1], Maarja Otsus [1] and Anne Kahru [1]

1 Laboratory of Environmental Toxicology, National Institute of Chemical Physics and Biophysics, 12618 Tallinn, Estonia; juri.laanoja@gmail.com (J.L.); sihtmae@gmail.com (M.S.); heiki.vija@kbfi.ee (H.V.); imbi.kurvet@kbfi.ee (I.K.); maarja.otsus@gmail.com (M.O.); anne.kahru@kbfi.ee (A.K.)

2 School of Science, Tallinn University of Technology, 19086 Tallinn, Estonia

* Correspondence: kaja.kasemets@kbfi.ee

† Presented at the International Conference EcoBalt 2023 "Chemicals & Environment", Tallinn, Estonia, 9–11 October 2023.

Keywords: wound infections; antimicrobial resistance; bacteria; fungi; novel antimicrobials; silver; chitosan; nanocomposites; synergy; particle-cell interactions

Hospital-acquired infections are serious medical problems worldwide. Therefore, novel antimicrobials for the treatment of infections, especially those caused by antibiotic-resistant microbes, are urgently needed. We have previously shown that Ag, CuO and ZnO nanoparticles are also toxic against pathogenic microbes and relatively safe to animal cells [1], thus being promising for medical use, e.g., in wound treatment. Moreover, combining these NPs with biologically active polymers may enhance their efficacy and specificity. Chitosan (CS) is a biocompatible, antimicrobial and immuno-modulating polymer and is already used for wound treatments. Therefore, crosslinking chitosan with antimicrobial nanoparticles can yield novel antimicrobials with both biocidal and immune-modulating effects. The study aimed (i) to design and synthesize silver–chitosan nanocomposites (nAgCSs) with different silver–chitosan (Ag/CS) weight ratios (1:0.3, 1:1 and 1:3), (ii) to evaluate their efficacy against bacteria and fungi that can cause wound infections and (iii) to elucidate the mode of antimicrobial action of nAgCSs. nAgCSs were synthesized through the reduction of $AgNO_3$ with trisodium citrate and stabilized/coated with low-molecular-weight chitosan. The antimicrobial activity of nAgCSs against bacteria *Pseudomonas aeruginosa*, *Escherichia coli* and *Staphylococcus aureus* and fungi *Candida albicans* and *C. glabrata* was studied using a Spot test [2]. In this test, microbes are exposed to toxicants in deionized water for 1, 4 and 24 h and then plated on an agar medium for the quantification of the minimum biocidal concentration (MBC). The synthesized nAgCSs' primary and hydrodynamic sizes were ~50 and ~100 nm, respectively, and the surface charge was ~+25 mV. The shedding of Ag ions was in the range of 2–4%. The synthesized nAgCSs were efficient antimicrobials acting already at sub-mg-per-litre concentrations. In general, the nAgCSs were more toxic towards bacteria than fungi (24-h MBC 0.07–0.56 and 9.3–44 mg Ag/L, respectively), and nAgCSs with an Ag/CS mass ratio of 1:3 were the most efficient. The high antimicrobial efficiency was most likely due to the absorption of nAgCSs onto the surface of the microbes, as shown via confocal laser scanning microscopy and flow cytometry. Interestingly, the shed Ag ions (the most reported toxicity mechanism of AgNPs) did not explain the biocidal effect of nAgCSs, indicating a synergy between chitosan and silver.

Citation: Kasemets, K.; Laanoja, J.; Sihtmäe, M.; Vija, H.; Kurvet, I.; Otsus, M.; Kahru, A. Silver–Chitosan Nanocomposites for Biomedical Application: Design, Synthesis and Antimicrobial Efficiency. *Proceedings* **2023**, *92*, 40. https://doi.org/10.3390/proceedings2023092040

Published: 24 November 2023

Author Contributions: Conceptualization, K.K. and A.K.; methodology, formal analysis and investigation, K.K.; J.L., M.S., H.V., I.K. and M.O.; writing, project administration and funding acquisition, K.K. All authors have read and agreed to the published version of the manuscript.

Funding: This research was funded by the Estonian Research Council project PRG 749 and conducted using the NAMUR+ core facility (TT13).

Institutional Review Board Statement: Not applicable.

Informed Consent Statement: Not applicable.

Data Availability Statement: Not applicable.

Conflicts of Interest: The authors declare no conflict of interest.

References

1. Suppi, S.; Kasemets, K.; Ivask, A.; Künnis-Beres, K.; Sihtmäe, M.; Kurvet, I.; Aruoja, V.; Kahru, A. A novel method for comparison of biocidal properties of nanomaterials to bacteria, yeasts and algae. *J. Hazard. Mater.* **2015**, *286*, 75–84. [CrossRef] [PubMed]
2. Bondarenko, O.; Juganson, K.; Ivask, A.; Kasemets, K.; Mortimer, M.; Kahru, A. Toxicity of Ag, CuO and ZnO nanoparticles to selected environmentally relevant test organisms and mammalian cells in vitro: A critical review. *Arch. Toxicol.* **2013**, *84*, 1181–1200. [CrossRef] [PubMed]

proceedings

Abstract

Polymeric Hydrogels for the Removal of Emerging Contaminants from Water †

Giusy Curcuruto [1], Chiara Zagni [2] and Sabrina Carola Carroccio [1,*]

[1] CNR-IPCB, Catania Unit, Via P. Gaifami 18, 95126 Catania, Italy; giusy.curcuruto@cnr.it
[2] Department of Drug and Health Sciences, University of Catania, Viale Andrea Doria 6, 95125 Catania, Italy; chiara.zagni@unict.it
* Correspondence: sabrinacarola.carroccio@cnr.it

† Presented at the International Conference EcoBalt 2023 "Chemicals & Environment", Tallinn, Estonia, 9–11 October 2023.

Abstract: Herein, a proof of concept is reported, concerning the formulation of methacrylic and styrene-based hydrogels with adsorption properties versus the emerging pollutants (dyes, pesticides, and pharmaceuticals). The synthesis of the hydrogels was achieved by functionalizing the HEMA monomer with specifically chelating groups, such as amino acids (lysine and histidine), cyclodextrins, and meglumine. The as-prepared monomers were polymerized in water at different temperatures to obtain macroporous samples. All the samples were characterized by chemical–physical and morphological analyses confirming the success of the reactions. The resulting systems were successfully tested to adsorb 2, 4 D, methylene blue, and lomefloxacin. In addition, up to five cycles of regeneration tests were performed, confirming the aptitude of the samples to be used several times without losing efficiency and maintaining their mechanical properties.

Keywords: wastewater; purification; hydrogel; photocatalysis; emerging contaminants

Citation: Curcuruto, G.; Zagni, C.; Carroccio, S.C. Polymeric Hydrogels for the Removal of Emerging Contaminants from Water. *Proceedings* **2023**, 92, 14. https://doi.org/10.3390/proceedings2023092014

Academic Editors: Monika Mortimer, Anne Kahru, Ivo Leito, Riin Rebane and Villem Aruoja

Published: 21 November 2023

Author Contributions: Conceptualization, S.C.C., and C.Z.; methodology, G.C. and C.Z., validation, G.C. and C.Z., investigation, G.C. and C.Z.; data curation, S.C.C.; writing—original draft preparation, S.C.C.; writing—review and editing S.C.C.; funding acquisition, S.C.C. All authors have read and agreed to the published version of the manuscript.

Funding: This work was partially founded by ANTIBIO-Antibiotics Removal From Water By Imprinted Magnetic Nanomaterials project, ProgettidiRicerca@CNR call 2020 (CUP: B63C22000010005).

Institutional Review Board Statement: Not applicable.

Informed Consent Statement: Not applicable.

Data Availability Statement: The data presented in this study are available in: doi: 10.1016/j.eurpolymj.2022.111556.

Acknowledgments: The authors wish to thank Emanuele Francesco Mirabella (CNR-IPCB) for the technical assistance.

Conflicts of Interest: The authors declare no conflict of interest.

proceedings

Abstract

Evaluation of Environmental Hazards of Nanocomposite Use for Wastewater Treatment †

Villem Aruoja [1,*], Giuliana Impellizzeri [2], Sabrina Carola Carroccio [3], Mariliis Sihtmäe [1], Irina Blinova [1] and Anne Kahru [1,4]

1 Laboratory of Environmental Toxicology, National Institute of Chemical Physics and Biophysics, 12618 Tallinn, Estonia; mariliis.sihtmae@kbfi.ee (M.S.); irina.blinova@kbfi.ee (I.B.); anne.kahru@kbfi.ee (A.K.)
2 CNR-IMM, 95123 Catania, Italy; giuliana.impellizzeri@ct.infn.it
3 CNR-IPCB, 95126 Catania, Italy; sabrinacarola.carroccio@cnr.it
4 Estonian Academy of Sciences, 10130 Tallinn, Estonia
* Correspondence: villem.aruoja@kbfi.ee
† Presented at the International Conference EcoBalt 2023 "Chemicals & Environment", Tallinn, Estonia, 9–11 October 2023.

Keywords: nanomaterials; environmental toxicology; antibiotic; wastewater treatment

Antibiotics are not completely metabolized nor degraded in traditional wastewater treatment, leaving a large proportion of them in the wastewater. Consequently, water bodies become reservoirs of antibiotics, where antibiotic resistance can develop. Therefore, a new strategy for the efficient, selective, and cost-effective removal of antibiotics from wastewater is being developed within the ANTIBIO project [1] using imprinted magnetic nanomaterials. However, in order to apply the new materials on a large scale, their potential harm to the environment has to be evaluated. We used aquatic ecotoxicology tests according to OECD/ISO to evaluate the environmental hazard of the novel nanocomposites. We used bacteria *Vibrio fischeri*, water fleas *Daphnia magna* and algae *Raphidocelis subcapitata* to study the toxicity of the nanocomposites as well as their component materials, including organic polymers, TiO_2, ZnO and magnetite (Fe_3O_4). The studied concentration range was 0.01–100 mg/L. Out of the tested samples, only ZnO was toxic to all organisms, with EC_{50} values (mg/L) 3.9 for *V. fischeri*, 3.3 for *D. magna* and 0.1 for *R. subcapitata*. When the inhibition was calculated based on mg Zn per liter, ZnO and $ZnSO_4$ were equally toxic, indicating the shedding of Zn ions as a possible mechanism of ZnO toxicity. Magnetite was toxic to algae, with an EC_{50} value of 2.2 mg/L; however, the composite containing magnetite core was not harmful to algae. In contrast to our previous studies [2–4], TiO_2 did not inhibit algal growth at 100 mg/L and the entrapment of algal cells within TiO_2 aggregates was not witnessed. According to our results, the organic polymers as well as the TiO_2 produced within this study are not environmentally harmful, while the toxicity of ZnO has to be considered when designing nanocomposites.

Author Contributions: Conceptualization, A.K., G.I. and S.C.C.; methodology, V.A., M.S. and I.B.; writing—original draft preparation, V.A.; writing—review and editing, V.A. and M.S.; visualization V.A. and M.S.; project administration, funding acquisition, A.K. All authors have read and agreed to the published version of the manuscript.

Funding: This research was funded by ANTIBIO—Antibiotics Removal From Water By Imprinted Magnetic Nanomaterials project, ProgettidiRicerca@CNR call 2020 (CUP:B63C22000010005), Principal investigator Giuliana Impellizzeri.

Institutional Review Board Statement: Not applicable.

Informed Consent Statement: Not applicable.

Citation: Aruoja, V.; Impellizzeri, G.; Carroccio, S.C.; Sihtmäe, M.; Blinova, I.; Kahru, A. Evaluation of Environmental Hazards of Nanocomposite Use for Wastewater Treatment. *Proceedings* **2023**, *92*, 71. https://doi.org/10.3390/proceedings2023092071

Published: 5 December 2023

Data Availability Statement: Data sharing is not applicable.

Conflicts of Interest: The authors declare no conflict of interest.

References

1. ANTIBIO. Availabe online: https://www.antibio.it/ (accessed on 5 October 2023).
2. Aruoja, V.; Pokhrel, S.; Sihtmae, M.; Mortimer, M.; Madler, L.; Kahru, A. Toxicity of 12 metal-based nanoparticles to algae, bacteria and protozoa. *Environ. Sci. Nano* **2015**, *2*, 630–644. [CrossRef]
3. Bondarenko, O.M.; Heinlaan, M.; Sihtmäe, M.; Ivask, A.; Kurvet, I.; Joonas, E.; Jemec, A.; Mannerström, M.; Heinonen, T.; Rekulapelly, R.; et al. Multilaboratory evaluation of 15 bioassays for (eco)toxicity screening and hazard ranking of engineered nanomaterials: FP7 project NANOVALID. *Nanotoxicology* **2016**, *10*, 1229–1242. [CrossRef] [PubMed]
4. Aruoja, V.; Dubourguier, H.C.; Kasemets, K.; Kahru, A. Toxicity of nanoparticles of CuO, ZnO and TiO2 to microalgae Pseudokirchneriella subcapitata. *Sci. Total Environ.* **2009**, *407*, 1461–1468. [CrossRef] [PubMed]

proceedings

MDPI

Abstract

Plasma Electrolytic Oxidation Synthesis of Heterostructured TiO_2 for Photoanode Applications †

Ramūnas Levinas [1,*], Vidas Pakštas [1], Algirdas Selskis [1], Tomas Murauskas [2], Roman Viter [3], Aldona Jagminienė [1], Ina Stankevičienė [1] and Eugenijus Norkus [1]

1 State Research Institute, Center for Physical Sciences and Technology (FTMC), Saulėtekio Ave. 3, LT-10257 Vilnius, Lithuania
2 Faculty of Chemistry and Geosciences, Vilnius University, Naugarduko Str. 24, LT-03225 Vilnius, Lithuania
3 Institute of Atomic Physics and Spectroscopy, University of Latvia, 19 Raina Blvd., LV-1586 Riga, Latvia
* Correspondence: ramunas.levinas@ftmc.lt
† Presented at the International Conference EcoBalt 2023 "Chemicals & Environment", Tallinn, Estonia, 9–11 October 2023.

Abstract: In the renewable energy field, the conversion of solar light into electrical or chemical energy is considered essential to moving towards a truly green energy economy. Solar energy can be harnessed not just through generating electricity with photovoltaic cells but also by driving photoelectrochemical (PEC) reactions such as water splitting or pollutant oxidation. In this study, TiO_2 films were synthesized electrochemically through a procedure called plasma electrolytic oxidation (PEO). Under specific conditions, as the Ti substrate dissolves and the oxide film grows, electron discharges occur across the film, and this ionizes both the oxide and some amount of electrolyte that had been in contact with it. The mixture then cools, leaving a macroporous TiO_2 structure. What is particularly interesting for PEC applications is that the films can be crystalline and doped after synthesis. XRD analysis revealed that a TiO_2 film that had been obtained at a voltage of 200 V had an anatase crystal structure. In addition, during ionization and cooling, ions from the solution can be incorporated into the film. By adding 0.1 M Cu_2SO_4 into the synthesis electrolyte, we were able to incorporate Cu into the films, as proven EDX and XPS. The TiO_2 and heterostructured films showed good PEC water-splitting activity and stability in alkaline media when illuminated with 365 nm LED light. It was found that the photocurrent obtained depends on the synthesis voltage and that the heterostructured films would generate ~2 times larger photocurrents. In addition, further surface functionalization (e.g., with Au) was investigated. Electron–hole recombination was evaluated using an advanced non-stationary photoelectrochemical technique—intensity-modulated photocurrent spectroscopy (IMPS). Generally, films have very little recombination and only at lower overpotentials up to ~1 V. Overall, the synthesis of oxide films through PEO may provide an efficient alternative to obtaining crystalline films via annealing, and various heterostructures can be created simply by modifying synthesis conditions.

Keywords: titanium oxide; plasma electrolytic oxidation; heterostructures; photoelectrochemistry; photoanode; water splitting

Citation: Levinas, R.; Pakštas, V.; Selskis, A.; Murauskas, T.; Viter, R.; Jagminienė, A.; Stankevičienė, I.; Norkus, E. Plasma Electrolytic Oxidation Synthesis of Heterostructured TiO_2 for Photoanode Applications. *Proceedings* **2023**, *92*, 51. https://doi.org/10.3390/proceedings2023092051

Academic Editors: Monika Mortimer, Anne Kahru, Ivo Leito, Riin Rebane and Villem Aruoja

Published: 24 November 2023

Author Contributions: Conceptualization, R.L. and E.N.; methodology, R.L. and E.N.; formal analysis, R.L.; investigation, R.L., V.P., A.S., T.M., R.V., A.J. and I.S.; resources, E.N.; writing—original draft preparation, R.L.; writing—review and editing, R.L. and E.N.; visualization, R.L.; supervision, E.N. All authors have read and agreed to the published version of the manuscript.

Funding: This research received funding from the Research Council of Lithuania (LMTLT), agreement no. S-PD-22-5—TICAL.

Institutional Review Board Statement: Not applicable.

Informed Consent Statement: Not applicable.

Data Availability Statement: Not applicable.

Conflicts of Interest: The authors declare no conflict of interest.

proceedings

Abstract

Spray-Pyrolysis Synthesised TiO_2 Thin Films for Photocatalytic Air Treatment from Volatile Organic Compounds †

Jekaterina Sydorenko *, Malle Krunks, Arvo Mere, Marina Krichevskaya and Ilona Oja Acik

Department of Materials and Environmental Technology, Tallinn University of Technology, Ehitajate tee 5, 19086 Tallinn, Estonia; malle.krunks@taltech.ee (M.K.); arvo.mere@taltech.ee (A.M.); marina.kritsevskaja@taltech.ee (M.K.); ilona.oja@taltech.ee (I.O.A.)

* Correspondence: jekaterina.spiridono@taltech.ee; Tel.: +372-620-3369

† Presented at the International Conference EcoBalt 2023 "Chemicals & Environment", Tallinn, Estonia, 9–11 October 2023.

Keywords: spray pyrolysis; TiO_2; thin films; photocatalysis; indoor air; purification; volatile organic compounds

A wide range of mixtures of volatile organic compounds (VOCs), which are present in indoor air in low concentrations, can strongly affect human health. Therefore, the scientific interest in photocatalytic oxidation as a cost-effective and efficient technology for the removal of VOCs from indoor air is growing. The aim of the study was to deposit TiO_2 thin films by ultrasonic spray pyrolysis with different titanium isopropoxide (TTIP)/acetylacetone (AcacH) molar ratios in the spray solution. Other objectives were to determine the optimal TTIP:AcacH ratio and to study the ability of the film to purify air from VOCs under different experimental conditions. TiO_2 films were deposited onto the borosilicate glass at 350 °C and heat-treated at 500 °C for 1 h. The TTIP:AcacH molar ratio in spray solution varied from 1:1 to 1:20. At first, the photocatalytic activity of all obtained films was estimated by the degradation of 8.8 mM stearic acid (SA) deposited on top of the film [1]. In the second step of the studies, TiO_2 films were tested for the oxidation of VOCs (acetone, acetaldehyde, heptane and toluene) as separate pollutants in the concentration ranges 5–40 ppm [2] and as 9 ppm VOCs mixtures [3]. The oxidation of VOCs was studied in the gas-phase, multi-section reactor under ultraviolet and visible light. An increase in the amount of AcacH in the spray solution enhanced the photocatalytic performance of the films due to the carbon incorporation and changes in electronic structure. The reaction rate constant of SA oxidation on the film with the TTIP:AcacH molar ratio 1:8 was 10 times higher than that of the 1:3 film. The TiO_2 film with a molar ratio of 1:8 showed a promising ability in VOCs' degradation, oxidizing up to 9 ppm VOCs mixtures at 1.5 min at catalyst surface areas of 600 cm^2 under ultraviolet, and up to 90% of the mixture under visible light.

Citation: Sydorenko, J.; Krunks, M.; Mere, A.; Krichevskaya, M.; Oja Acik, I. Spray-Pyrolysis Synthesised TiO_2 Thin Films for Photocatalytic Air Treatment from Volatile Organic Compounds. *Proceedings* **2023**, *92*, 37. https://doi.org/10.3390/proceedings2023092037

Academic Editors: Monika Mortimer, Anne Kahru, Ivo Leito, Riin Rebane and Villem Aruoja

Published: 23 November 2023

Author Contributions: Conceptualization, J.S., I.O.A. and M.K. (Marina Krichevskaya); methodology, I.O.A., M.K. (Marina Krichevskaya); software, A.M., M.K. (Marina Krichevskaya); validation, J.S.; formal analysis, J.S., I.O.A. and M.K. (Marina Krichevskaya); investigation, I.O.A., M.K. (Marina Krichevskaya), M.K. (Malle Krunks) and A.M.; resources, I.O.A.; data curation, J.S.; writing—original draft preparation, J.S.; writing—review and editing, I.O.A., M.K. (Marina Krichevskaya), M.K. (Malle Krunks) and A.M.; visualization, J.S.; supervision, I.O.A. and M.K. (Marina Krichevskaya); project administration, I.O.A.; funding acquisition, I.O.A. All authors have read and agreed to the published version of the manuscript.

Funding: This research was funded by The Estonian Ministry of Education and Research institutional research funding project PRG627 and the European Commission's H2020 programme under the ERA Chair project 5GSOLAR grant agreement No 952509.

Institutional Review Board Statement: Not applicable.

Informed Consent Statement: Not applicable.

Data Availability Statement: Not applicable.

Conflicts of Interest: The authors declare no conflict of interest.

References

1. Spiridonova, J.; Katerski, A.; Danilson, M.; Krichevskaya, M.; Krunks, M.; Acik, I.O. Effect of the Titanium Isopropoxide:Acetylacetone Molar Ratio on the Photocatalytic Activity of TiO_2 Thin Films. *Molecules* **2019**, *24*, 4326. [CrossRef] [PubMed]
2. Spiridonova, J.; Mere, A.; Krunks, M.; Rosenberg, M.; Kahru, A.; Danilson, M.; Krichevskaya, M.; Acik, I.O. Enhanced Visible and Ultraviolet Light-Induced Gas-Phase Photocatalytic Activity of TiO_2 Thin Films Modified by Increased Amount of Acetylacetone in Precursor Solution for Spray Pyrolysis. *Catalysis* **2020**, *10*, 1011. [CrossRef]
3. Sydorenko, J.; Mere, A.; Krunks, M.; Krichevskaya, M.; Acik, I.O. Transparent TiO_2 thin films with high photocatalytic activity for indoor air purification. *RCS Adv.* **2022**, *12*, 35531–35542. [CrossRef] [PubMed]

proceedings

MDPI

Abstract

Designing of Sustainable Building Material Made of Non-Fired Clay with Various Biopolymers †

Yahor Trambitski *, Olga Kizinievič and Viktor Kizinievič

Laboratory of Composite Materials, Institute of Building Materials, Vilnius Gediminas Technical University, Linkmenų g. 28, LT 08217 Vilnius, Lithuania; olga.kizinievic@vilniustech.lt (O.K.); viktor.kizinievic@vilniustech.lt (V.K.)

* Correspondence: yahor.trambitski@vilniustech.lt; Tel.: +370-64519205

† Presented at the International Conference EcoBalt 2023 "Chemicals & Environment", Tallinn, Estonia, 9–11 October 2023.

Keywords: clay; biopolymers; stabilization; modification; building materials; sustainability

This study is dedicated to the development of an environmentally friendly building material made from non-fired clay, stabilized with various biopolymers (starch, alginate, and chitosan). The efficacy of the application of natural polymers has already been proven in the geotechnical industry for soil stabilization [1–3]. The authors believe that the use of natural polymers will have a positive impact on the physical–mechanical and operational properties of the newly obtained building material made from non-fired clay. With the addition of different concentrations (2.5%, 5%, 7.5%, and 10%) of biopolymer solutions, the density of the clay composites was reduced by up to 9% (from 1990 to 1810 kg/m^3). The inclusion of starch and alginate contributed to an increase in the strength of the clay composite of up to 42% (from 6.6 to 9.4 MPa). However, the application of 10% chitosan resulted in a 20% decrease in strength (from 6.6 to 5.3 MPa). The conducted research confirms that biopolymers are a promising stabilizer for non-fired clay composites. Nevertheless, to create an effective building material, it is essential to consider the type of clay and biopolymer, as well as their interaction potential and compatibility.

Citation: Trambitski, Y.; Kizinievič, O.; Kizinievič, V. Designing of Sustainable Building Material Made of Non-Fired Clay with Various Biopolymers. *Proceedings* **2023**, *92*, 56. https://doi.org/10.3390/proceedings2023092056

Academic Editors: Monika Mortimer, Anne Kahru, Ivo Leito, Riin Rebane and Villem Aruoja

Published: 28 November 2023

Author Contributions: Conceptualization, Y.T. and O.K.; methodology, O.K. and V.K.; validation, Y.T., O.K. and V.K.; data curation, O.K.; writing—original draft preparation, Y.T.; writing—review and editing, O.K. and V.K.; visualization, Y.T.; supervision, O.K. All authors have read and agreed to the published version of the manuscript.

Funding: This research received no external funding.

Institutional Review Board Statement: Not applicable.

Informed Consent Statement: Not applicable.

Data Availability Statement: The data presented in this study are available on request from the corresponding author.

Conflicts of Interest: The authors declare no conflict of interest.

References

1. Chang, I.; Im, J.; Prasidhi, A.K.; Cho, G.C. Effects of Xanthan gum biopolymer on soil strengthening. *Constr. Build. Mater.* **2015**, *74*, 65–72. [CrossRef]
2. Losini, A.E.; Grillet, A.C.; Bellotto, M.; Woloszyn, M.; Dotelli, G. Natural additives and biopolymers for raw earth construction stabilization—A review. *Constr. Build. Mater.* **2021**, *304*, 124507. [CrossRef]
3. Soldo, A.; Miletić, M.; Auad, M.L. Biopolymers as a sustainable solution for the enhancement of soil mechanical properties. *Sci. Rep.* **2020**, *10*, 267. [CrossRef] [PubMed]

proceedings

Abstract

Metal–Phenolic Network-Coated Nanoparticles for Reducing the Toxicity of Metal Nanomaterials †

Monika Mortimer [1,*], Anett Välimets [1,2] and Anne Kahru [1]

1 Laboratory of Environmental Toxicology, National Institute of Chemical Physics and Biophysics, Akadeemia Tee 23, 12618 Tallinn, Estonia; anettvalimets@gmail.com (A.V.); anne.kahru@kbfi.ee (A.K.)
2 School of Natural Sciences and Health, Tallinn University, Narva mnt 25, 10120 Tallinn, Estonia
* Correspondence: monika.mortimer@kbfi.ee

† Presented at the International Conference EcoBalt 2023 "Chemicals & Environment", Tallinn, Estonia, 9–11 October 2023.

Keywords: iron; tannic acid; gold nanoparticles; biocompatibility; toxicity; metals; adsorbent; aquatic

The growing use of metal nanomaterials (including Ag-, Cu- and Zn-based nanoparticles (NPs)) in medical applications but also in various green technologies is expected to result in an elevated environmental burden of toxic metals and NPs [1]. Since soluble metal-based NPs cause toxicity via released toxic metal ions [2,3], the safe use of NPs requires strategies for mitigating their toxicity via the removal of these metal ions. Here, a novel class of nano-sized specific adsorbent materials, metal–phenolic networks, are proposed as suitable materials for this purpose due to their biocompatibility, high specific surface area and the presence of functional groups specific for metal-ion binding. Iron–tannic-acid-network-coated Au nanoparticles (Fe-TA@Au NPs) were synthesized and characterized for their physicochemical properties and metal adsorption profile, using Cu ions as model toxicants. The morphology, size, composition and stability in water of the synthesized adsorbent materials were characterized using electron microscopy, Fourier-transform infrared spectroscopy (FTIR), dynamic light scattering, ultraviolet–visible spectrophotometry (UV-Vis) and TXRF spectrometry (S2 PICOFOX, Bruker, Billerica, Massachusetts, USA). The metal ion adsorption capacity, kinetics and specificity of the synthesized Fe-TA@Au NPs were determined in aqueous solutions containing Cu^{2+} ions. A facile two-step synthesis in an aqueous medium at room temperature yielded TA-stabilized Au NPs with a primary size of 25 ± 7 nm which were coated with an Fe-TA amorphous layer (thickness: 7.6 ± 3 nm). The hydrodynamic diameter of the Fe-TA@Au NPs was ~60 nm, and the surface charge was highly negative in both MilliQ water (pH 6.0) and in HEPES buffer (pH of 7.4; zeta potential of −45 and −60 mV, respectively). Aqueous suspensions of Fe-TA@Au NPs were stable over several days. An FTIR analysis indicated the presence of metal coordination bonds between TA and Fe atoms in the metal–phenolic network which were essential for the formation of the network structure. The Fe-TA@Au NPs effectively adsorbed Cu^{2+} in aqueous media, as determined via TXRF spectrometry. When unicellular freshwater protozoa *Tetrahymena thermophila* [4–6] were co-incubated with Fe-TA@Au NPs and $CuSO_4$, the Fe-TA@Au NPs completely rescued protozoa from the toxicity of $CuSO_4$, suggesting efficient adsorption of the Cu ions by the synthesized metal–phenolic networks. The results indicate that NPs coated with metal–phenolic networks have promising applications in environmental remediation.

Author Contributions: Conceptualization, M.M.; methodology, M.M.; investigation, M.M. and A.V.; writing—original draft preparation, M.M.; writing—review and editing, M.M. and A.K.; project administration, M.M.; funding acquisition, M.M. All authors have read and agreed to the published version of the manuscript.

Citation: Mortimer, M.; Välimets, A.; Kahru, A. Metal–Phenolic Network-Coated Nanoparticles for Reducing the Toxicity of Metal Nanomaterials. *Proceedings* **2023**, *92*, 24. https://doi.org/10.3390/proceedings2023092024

Academic Editors: Ivo Leito, Riin Rebane and Villem Aruoja

Published: 21 November 2023

Funding: This research was funded by the Estonian Research Council, grant number STP28.

Institutional Review Board Statement: Not applicable.

Informed Consent Statement: Not applicable.

Data Availability Statement: Data supporting reported results will be made available upon request.

Acknowledgments: The authors thank Heiki Vija for assistance with chemical analysis and Maarja Otsus for assistance with microscopy.

Conflicts of Interest: The authors declare no conflict of interest.

References

1. Mortimer, M.; Kahru, A. Nanomaterials and Microorganisms: From Green Synthesis to Antibacterial Applications in Medicine and Agriculture. *Nanomaterials* **2022**, *12*, 4265. [CrossRef]
2. Ivask, A.; Juganson, K.; Bondarenko, O.; Mortimer, M.; Aruoja, V.; Kasemets, K.; Blinova, I.; Heinlaan, M.; Slaveykova, V.; Kahru, A. Mechanisms of toxic action of Ag, ZnO and CuO nanoparticles to selected ecotoxicological test organisms and mammalian cells in vitro: A comparative review. *Nanotoxicology* **2014**, *8* (Suppl. S1), 57–71. [CrossRef]
3. Guo, L.-H.; Mortimer, M. (Eds.) *Advances in Toxicology and Risk Assessment of Nanomaterials and Emerging Contaminants*; Springer Nature: Singapore, 2022; pp. 1–371. [CrossRef]
4. Mortimer, M.; Kasemets, K.; Kahru, A. Toxicity of ZnO and CuO nanoparticles to ciliated protozoa *Tetrahymena thermophila*. *Toxicology* **2010**, *269*, 182–189. [CrossRef]
5. Mortimer, M.; Kasemets, K.; Vodovnik, M.; Marinsek-Logar, R.; Kahru, A. Exposure to CuO nanoparticles changes the fatty acid composition of protozoa *Tetrahymena thermophila*. *Environ. Sci. Technol.* **2011**, *45*, 6617–6624. [CrossRef]
6. Juganson, K.; Mortimer, M.; Ivask, A.; Pucciarelli, S.; Miceli, C.; Orupold, K.; Kahru, A. Mechanisms of toxic action of silver nanoparticles in the protozoan *Tetrahymena thermophila*: From gene expression to phenotypic events. *Environ. Pollut.* **2017**, *225*, 481–489. [CrossRef]

proceedings

MDPI

Abstract

Machine Learning Tools Can Pinpoint High-Risk Water Pollutants †

Helen Sepman [1,2,‡], Pilleriin Peets [1,3], Lisa Jonsson [1], Louise Malm [1], Malte Posselt [2], Matthew MacLeod [2], Jonathan Martin [2], Magnus Breitholtz [2], Michael McLachlan [2] and Anneli Kruve [1,2,*]

1 Department of Materials and Environmental Chemistry, Stockholm University, Svante Arrhenius väg 16, 106 91 Stockholm, Sweden; helen.sepman@aces.su.se (H.S.)
2 Department of Environmental Science, Stockholm University, Svante Arrhenius väg 8, 106 91 Stockholm, Sweden
3 Institute of Biodiversity, Faculty of Biological Science, Cluster of Excellence Balance of the Microverse, Friedrich Schiller University Jena, 07743 Jena, Germany
* Correspondence: anneli.kruve@su.se

† Presented at the International Conference EcoBalt 2023 "Chemicals & Environment", Tallinn, Estonia, 9–11 October 2023.
‡ This author is Presenting author.

Keywords: high-resolution mass spectrometry; toxicity; quantification; machine learning; non-targeted screening; suspect screening; risk assessment

Citation: Sepman, H.; Peets, P.; Jonsson, L.; Malm, L.; Posselt, M.; MacLeod, M.; Martin, J.; Breitholtz, M.; McLachlan, M.; Kruve, A. Machine Learning Tools Can Pinpoint High-Risk Water Pollutants. *Proceedings* **2023**, *92*, 68. https://doi.org/10.3390/proceedings2023092068

Published: 30 November 2023

Liquid chromatography–high-resolution mass spectrometry (LC/HRMS) is a powerful tool for detecting chemicals that are present in low concentrations. While this technique has revealed thousands of ionizable pollutants in environmental samples [1,2], the expanding list of emerging contaminants highlights the urgency to speed up their risk assessment [3,4]. Generally, the risk assessment workflow starts with structural identification, followed by obtaining the analytical standard for confirmation, toxicity assessment, and quantification with a calibration curve. To speed up the process, machine learning has found use in predicting toxicity and ionization efficiency; however, most of the models in use still require a chemical structure as an input. Therefore, detected but unidentified chemicals are frequently discarded from further analysis and the bioactivity of samples often remains partially unexplained [5]. Still, the fragmentation spectrum provides information about the structure which can be related to the properties of the chemical. We developed a workflow for estimating the risk of chemicals detected in non-target screening based on their MS^2 data. Two prediction models, MS2Quant [6] for ionization efficiency and MS2Tox [7] for acute fish toxicity, were trained based on structural fingerprints. While structural fingerprints can be calculated from a structure, the recently developed SIRIUS+CSI:FingerID software [8] offers the possibility to predict these fingerprints based on the MS^2 spectrum, and therefore predict chemical properties without structural assignment. Based on the validation set, the root mean square errors of MS2Quant and MS2Tox were 5.9× (39 chemicals) and 7.8× (219 chemicals), respectively. These models were applied in a non-target screening workflow regarding wastewater analysis. The preliminary results show that MS2Quant and MS2Tox help to pinpoint chemicals that pose a higher risk compared to a top five approach. Therefore, this approach provides the possibility to evaluate the risk of unidentified LC/HRMS features and prioritize high-risk chemicals in identification.

Author Contributions: H.S., P.P. and A.K. designed the research study. H.S. and P.P. developed the models and wrote the code. L.J., H.S. and L.M. performed the measurements. A.K., M.P. and M.M. (Michael McLachlan) performed supervision. A.K., J.M., M.M. (Matthew MacLeod) and M.B. acquired funding for the project. All authors have read and agreed to the published version of the manuscript.

Funding: The funding was generously provided by the Swedish Research Council for Sustainable Development, grant 2020-01511.

Institutional Review Board Statement: Not applicable.

Informed Consent Statement: Not applicable.

Data Availability Statement: The data will be made available together with the full publication.

Acknowledgments: The authors would like to thank Claudia Möckel and Merle Plassmann for their technical support.

Conflicts of Interest: The authors declare no competing financial interests.

References

1. Papazian, S.; D'Agostino, L.A.; Sadiktsis, I.; Froment, J.; Bonnefille, B.; Sdougkou, K.; Xie, H.; Athanassiadis, I.; Budhavant, K.; Dasari, S.; et al. Nontarget Mass Spectrometry and in Silico Molecular Characterization of Air Pollution from the Indian Subcontinent. *Commun. Earth Environ.* **2022**, *3*, 35. [CrossRef]
2. Schymanski, E.L.; Singer, H.P.; Slobodnik, J.; Ipolyi, I.M.; Oswald, P.; Krauss, M.; Schulze, T.; Haglund, P.; Letzel, T.; Grosse, S.; et al. Non-Target Screening with High-Resolution Mass Spectrometry: Critical Review Using a Collaborative Trial on Water Analysis. *Anal. Bioanal. Chem.* **2015**, *407*, 6237–6255. [CrossRef] [PubMed]
3. Richardson, S.D.; Ternes, T.A. Water Analysis: Emerging Contaminants and Current Issues. *Anal. Chem.* **2022**, *94*, 382–416. [CrossRef] [PubMed]
4. McCord, J.P.; Groff, L.C.; Sobus, J.R. Quantitative Non-Targeted Analysis: Bridging the Gap between Contaminant Discovery and Risk Characterization. *Environ. Int.* **2022**, *158*, 107011. [CrossRef] [PubMed]
5. Houtman, C.J.; ten Broek, R.; van Oorschot, Y.; Kloes, D.; van der Oost, R.; Rosielle, M.; Lamoree, M.H. High Resolution Effect-Directed Analysis of Steroid Hormone (Ant)Agonists in Surface and Wastewater Quality Monitoring. *Environ. Toxicol. Pharmacol.* **2020**, *80*, 103460. [CrossRef] [PubMed]
6. Sepman, H.; Malm, L.; Peets, P.; MacLeod, M.; Martin, J.; Breitholtz, M.; Kruve, A. Bypassing the Identification: MS2Quant for Concentration Estimations of Chemicals Detected with Nontarget LC-HRMS from MS 2 Data. *Anal. Chem.* **2023**, *95*, 12329–12338. [CrossRef] [PubMed]
7. Peets, P.; Wang, W.-C.; MacLeod, M.; Breitholtz, M.; Martin, J.W.; Kruve, A. MS2Tox Machine Learning Tool for Predicting the Ecotoxicity of Unidentified Chemicals in Water by Nontarget LC-HRMS. *Environ. Sci. Technol.* **2022**, *56*, 15508–15517. [CrossRef]
8. Dührkop, K.; Fleischauer, M.; Ludwig, M.; Aksenov, A.A.; Melnik, A.V.; Meusel, M.; Dorrestein, P.C.; Rousu, J.; Böcker, S. SIRIUS 4: A Rapid Tool for Turning Tandem Mass Spectra into Metabolite Structure Information. *Nat. Methods* **2019**, *16*, 299–302. [CrossRef]

proceedings

Abstract

Microplastics in Influents and Effluents of Estonian Wastewater Treatment Plants †

Ayankoya Yemi Ayankunle [1,*], Natalja Buhhalko [2], Aljona Lukjanova [3], Karin Pachel [1], Erki Lember [1] and Margit Heinlaan [3,*]

1 Department of Environmental Engineering, Tallinn University of Technology, Ehitajate tee 5, 19086 Tallinn, Estonia; karin.pachel@taltech.ee (K.P.); erkilember@gmail.com (E.L.)
2 Department of Marine Systems, Tallinn University of Technology, Akadeemia tee 15A,12618 Tallinn, Estonia; natalja.buhhalko@taltech.ee
3 Laboratory of Environmental Toxicology, National Institute of Chemical Physics and Biophysics, Akadeemia tee 23, 12618 Tallinn, Estonia; aljona.lukjanova@kbfi.ee
* Correspondence: ayankoya.ayankunle@taltech.ee (A.Y.A.); margit.heinlaan@kbfi.ee (M.H.)
† Presented at the International Conference EcoBalt 2023 "Chemicals & Environment", Tallinn, Estonia, 9–11 October 2023.

Keywords: polymers; μFTIR; plastic pollution

This study is the first systematic investigation of microplastics in influents and effluents of Estonian wastewater treatment plants (WWTPs) using continuous sampling filtration method and FTIR-spectroscopy. The aim of the study was to evaluate the efficiency of WWTPs in removing microplastics from the treated water. In total, six WWTPs from all over Estonia were included in the study. For sampling, 14, 25 and 1450 L samples of wastewater were collected from the influents before and after the screen and effluents, respectively, and filtrated over 24 h on three layers (5, 0.4, 0.33 mm) of metal sieves. For microplastics analysis, organics in the sample were oxidized and the samples were transferred to 10 μm polycarbonate filters for microscope-aided characterization. Selected particles were identified using μFTIR. Contamination controls from sample collection, preparation and characterization steps were analyzed in parallel. The results showed that microplastics were present in all the samples, with the highest concentrations observed in the influent sampled before the screen. The most common polymer types identified were polyester, polyethylene and polypropylene. The study found that the treatment process was effective at removing larger-sized microplastic fractions but less effective in removing smaller ones. The obtained data are also important for estimating the microplastics load that reaches the environment in WWTP sludge used in agriculture and landscaping. According to Koelmans et al. (2019) [1], the results of the study can be considered of high quality and are hence important for implementing microplastics mitigation strategies and control. The study established baseline levels of microplastics in the influent and effluent of Estonian WWTPs.

Citation: Ayankunle, A.Y.; Buhhalko, N.; Lukjanova, A.; Pachel, K.; Lember, E.; Heinlaan, M. Microplastics in Influents and Effluents of Estonian Wastewater Treatment Plants. *Proceedings* **2023**, *92*, 55. https://doi.org/10.3390/proceedings2023092055

Published: 27 November 2023

Author Contributions: Conceptualization, A.Y.A. and M.H.; methodology, A.Y.A., M.H. and N.B.; software, A.Y.A., E.L., A.L. and M.H.; validation, A.Y.A., M.H., A.L. and N.B.; formal analysis, A.Y.A. and M.H.; investigation, A.Y.A., M.H., A.L. and N.B.; resources, A.Y.A., M.H., E.L. and K.P.; data curation, A.Y.A., E.L., N.B. and M.H.; writing—original draft preparation, A.Y.A. and M.H.; writing—review and editing, A.Y.A., E.L. and M.H.; visualization, A.Y.A., E.L. and M.H.; supervision, M.H., E.L. and K.P.; project administration, M.H., E.L. and K.P.; funding acquisition, M.H. and K.P. All authors have read and agreed to the published version of the manuscript.

Funding: This research was funded by the Estonian Research Council grant number PRG1427 (A.Y.A., A.L. and M.H.) and AS Emajõe Veevärk.

Institutional Review Board Statement: Not applicable.

Informed Consent Statement: Not applicable.

Data Availability Statement: Data is unavailable due to privacy.

Acknowledgments: Authors are thankful to the staff of the investigated wastewater treatment plant in Estonia for their support and hospitality during sample collection. Also, the staff and colleagues from AS Emajõe Veevärk are acknowledged for their administrative and technical support.

Conflicts of Interest: The authors declare no conflict of interest.

Reference

1. Koelmans, A.A.; Mohamed Nor, N.H.; Hermsen, E.; Kooi, M.; Mintenig, S.M.; De France, J. Microplastics in freshwaters and drinking water: Critical review and assessment of data quality. *Water Res.* **2019**, *155*, 410–422. [CrossRef] [PubMed]

proceedings

MDPI

Abstract

The Water Flea as a "Canary in the Coal Mine"—Using Phenotypic and Molecular Endpoints to Understand Pollution †

Emma Rowan, Anne Leung, Maria Giannouli, Katie O'Rourke and Konstantinos Grintzalis *

School of Biotechnology, Dublin City University, D09 DX63 Dublin, Ireland; emma.rowan@dcu.ie (E.R.); anne.leung@dcu.ie (A.L.); maria.giannouli2@mail.dcu.ie (M.G.); katie.orourke9@mail.dcu.ie (K.O.)
* Correspondence: konstantinos.gkrintzalis@dcu.ie
† Presented at the International Conference EcoBalt 2023 "Chemicals & Environment", Tallinn, Estonia, 9–11 October 2023.

Abstract: The assessment of pollution is a serious issue and a major consequence of the overgrowing human population and its activities. Focusing on the aquatic ecosystem, traditional approaches of water chemistry mainly provide minimal monitoring with the detection of pollutants, while they fail to produce mechanistic or predictive insight. As such, effect-based methods have gained significant attention for the better mechanistic understanding of aquatic pollution. Among the key species used, daphnids have acquired a central position in aquatic toxicology and ecology. In this study, a novel feeding assay was developed and applied in a battery of exposures to different pollutants. Furthermore, in combination with biochemical markers and sensitive metabolomic analyses, the responses of daphnids following exposures were uncovered in molecular detail. Specific categories of metabolites were identified as significant indicators to predict pollution.

Keywords: molecular ecotoxicology; *Daphnia*; metabolomics; feeding; pharmaceuticals; pollutants

Citation: Rowan, E.; Leung, A.; Giannouli, M.; O'Rourke, K.; Grintzalis, K. The Water Flea as a "Canary in the Coal Mine"—Using Phenotypic and Molecular Endpoints to Understand Pollution. *Proceedings* **2023**, 92, 12. https://doi.org/10.3390/proceedings2023092012

Academic Editors: Monika Mortimer, Anne Kahru, Ivo Leito, Riin Rebane and Villem Aruoja

Published: 21 November 2023

Author Contributions: Conceptualization, resources, writing—original draft preparation, writing—review and editing, supervision, funding acquisition, K.G.; investigation, data acquisition, E.R., A.L., M.G. and K.O. All authors have read and agreed to the published version of the manuscript.

Funding: This research was funded by SCIENCE FOUNDATION IRELAND under grant number (18/SIRG/5563 Metabolomic approaches in mechanistic toxicology) and the IRISH RESEARCH COUNCIL in support of Katie O'Rourke under grant number (GOIPG/2020/199 Integration of holistic approaches to detect pharmaceuticals in the aquatic environment).

Institutional Review Board Statement: Ethical review and approval were waived for this study, due to the fact that daphnids are regarded as "animals" in terms of being members of the kingdom Animalia, however, they are not "animals" as defined in regulation SI543 of 2012 on the protection of animals used for scientific purposes. Therefore, the study does not require authorization from the Health Products Regulatory Authority (HPRA), while is also in line with the aim of working under the 3Rs (reduce, refine, replacement) strategy, since daphnids are commonly used in ecology and ecotoxicology as replacements of more evolutionary advanced species (i.e., fishes), posing no ethical implications.

Informed Consent Statement: Not applicable.

Data Availability Statement: Data sharing is not applicable to this abstract.

Acknowledgments: The authors acknowledge the support from the Metabolomics Platform of the University College Dublin and the University of Heidelberg.

Conflicts of Interest: The authors declare no conflict of interest. The funders had no role in the design of the study; in the collection, analyses, or interpretation of data; in the writing of the manuscript; or in the decision to publish the results.

proceedings

Abstract

Use of Midge *Chironomus riparius* Larvae in Plastic Ecotoxicity Studies and Peculiarities of Their Responses †

Alla Khosrovyan [1,*] and Anne Kahru [1,2]

1 Laboratory of Environmental Toxicology, National Institute of Chemical Physics and Biophysics, 23 Akadeemia Tee, 12618 Tallinn, Estonia; anne.kahru@kbfi.ee
2 Estonian Academy of Sciences, 6 Kohtu, 10130 Tallinn, Estonia
* Correspondence: alla.khosrovyan@kbfi.ee

† Presented at the International Conference EcoBalt 2023 "Chemicals & Environment", Tallinn, Estonia, 9–11 October 2023.

Keywords: multi-generation; UV-weathering

Citation: Khosrovyan, A.; Kahru, A. Use of Midge *Chironomus riparius* Larvae in Plastic Ecotoxicity Studies and Peculiarities of Their Responses. *Proceedings* **2023**, *92*, 63. https://doi.org/10.3390/proceedings2023092063

Academic Editors: Monika Mortimer, Ivo Leito, Riin Rebane and Villem Aruoja

Published: 29 November 2023

Chironomus riparius, a common test organism [1], has been widely used for the assessment of the ecotoxicity of metal and organic contaminants in aquatic systems. Due to having both aquatic and sediment-based developmental stages, midge larvae can be exposed to both waterborne and sediment-bound contaminants, and hence, represent a versatile model for toxicity studies. Recently, *C. riparius* has also been widely used in the environmental risk assessment of microplastic contamination in water bodies. The responses of chironomid larvae that were experimentally exposed to nano- or microplastic particles were shown to range from the molecular to whole organism level, e.g., [2–4]. However, these responses have not been strongly related to the microplastic concentrations used for exposure or load within the organism (the number of ingested particles). Instead, although the presence of microplastic particles in the gut of *C. riparius* was demonstrated [5,6], the populations of *C. riparius* that were continuously exposed to microplastic during three generations showed recovery after the first generation [7]. Moreover, despite the greater number of ingested particles at a higher microplastic concentration (1 g kg^{-1}), compared to the 10 times lower concentration (0.1 g kg^{-1}), the emergence of adult midges in both cases was not significantly different [6]. In this work, a review of the responses of *C. riparius* to plastic exposure based on the published literature and our own data is provided, and the peculiarities of the observed responses are discussed.

Author Contributions: Conceptualization, A.K. (Alla Khosrovyan); methodology, A.K. (Alla Khosrovyan) and A.K. (Anne Kahru); formal analysis, A.K. (Alla Khosrovyan); investigation, A.K. (Alla Khosrovyan); resources, A.K. (Anne Kahru); writing—original draft preparation, A.K. (Alla Khosrovyan); writing—review and editing, A.K. (Anne Kahru); project administration, A.K. (Alla Khosrovyan) and A.K. (Anne Kahru); funding acquisition, A.K. (Alla Khosrovyan) and A.K. (Anne Kahru). All authors have read and agreed to the published version of the manuscript.

Funding: This research was funded by Estonian Research Council grants (Mobilitas Pluss 445 MOBJD509, TT13 and PRG749), European Regional Development Fund grants (NAMUR+ 2014- 446 2020.4.01.16-0123 and TK134) and an institutional grant (Arengufond_AK).

Institutional Review Board Statement: Not applicable.

Informed Consent Statement: Not applicable.

Data Availability Statement: No new data were created or analyzed in this study. Data sharing is not applicable to this article.

Conflicts of Interest: The authors declare no conflict of interest.

References

1. Organization for Economic CO-Operation and Development (OECD). Sediment-Water Chironomid Life-Cycle Toxicity Test Using Spiked Water or Spiked Sediment. Test N 233. 2010. Available online: https://www.oecd.org/env/test-no-233-sediment-water-chironomid-life-cycle-toxicity-test-using-spiked-water-or-spiked-sediment-9789264090910-en.htm (accessed on 8 October 2023).
2. Silva, C.J.M.; Silva, A.L.P.; Gravato, C.; Pestana, J.L.T. Ingestion of small-sized and irregularly shaped polyethylene microplastics affect Chironomus riparius life-history traits. *Sci. Total Environ.* **2019**, *672*, 862–868. [CrossRef] [PubMed]
3. Khosrovyan, A.; Doria, H.B.; Kahru, A.; Pfenninger, M. Polyamide microplastic exposure elicits rapid, strong and genome-wide evolutionary response in the freshwater non-biting midge Chironomus riparius. *Chemosphere* **2022**, *299*, 134452. [CrossRef] [PubMed]
4. Almroth, B.C.; Carle, A.; Blanchard, M.; Molinari, F.; Bour, A. Single-use take-away cups of paper are as toxic to aquatic midge larvae as plastic cups. *Environ. Pollut.* **2023**, *830*, 121836. [CrossRef] [PubMed]
5. Scherer, C.; Brennholt, N.; Reifferscheid, G.; Wagner, M. Feeding type and development drive the ingestion of microplastics by freshwater invertebrates. *Sci. Rep.* **2017**, *7*, 17006. [CrossRef] [PubMed]
6. Khosrovyan, A.; Gabrielyan, B.; Kahru, A. Ingestion and effects of virgin polyamide microplastics on Chironomus riparius adult larvae and adult zebrafish Danio rerio. *Chemosphere* **2020**, *259*, 127456. [CrossRef] [PubMed]
7. Khosrovyan, A.; Kahru, A. Evaluation of the potential toxicity of UV-weathered virgin polyamide microplastics to non-biting midge Chironomus riparius. *Environ. Pollut.* **2021**, *287*, 117334. [CrossRef] [PubMed]

proceedings

MDPI

Abstract

Production and Characterization of Polysaccharides from *Rhodotorula toruloides* †

Henrique Sepulveda Del Rio Hamacek *, Petri-Jaan Lahtvee and Rahul Kumar

Department of Chemistry and Biotechnology, Tallinn University of Technology, Tallinn 12618, Estonia; petrijaan.lahtvee@taltech.ee (P.-J.L.); rkumar@taltech.ee (R.K.)

* Correspondence: henrique.sepulveda@taltech.ee

† Presented at the International Conference EcoBalt 2023 "Chemicals & Environment", Tallinn, Estonia, 9–11 October 2023.

Keywords: yeast; polysaccharides; sustainable material production; *Rhodotorula toruloides*

The increasing demand for sustainable material production has fueled extensive research on alternative polysaccharide sources. Microbial polysaccharides, particularly exopolysaccharides (EPSs) in yeast, have garnered significant attention in the industrial sector for their unique properties and production methods [1]. *Rhodotorula toruloides*, an unconventional yeast known for intracellular lipids and carotenoid production, also possesses the capacity to produce EPSs [2]. EPS are primarily composed of mannose and glucose, alongside non-carbohydrate components such as proteins and glycoproteins [3]. Therefore, this study aimed to investigate, characterize, and evaluate the chemical properties of EPSs derived from *R. toruloides*. The yeast was cultivated in shaker flasks under varied growth conditions, including various pH levels, salt concentrations, agitation speeds, and C/N ratios, allowing the identification of the optimal environment for EPS production. Previous research has shown that a high carbon-to-limiting nutrient ratio in the growth media increases EPS production [4]. Additionally, an increase in the ionic strength of the medium hinders *R. toruloides* flocculation [5]. EPS purification was achieved through absolute ethanol precipitation, followed by characterization via HPLC for monosaccharide composition, GC-MS for glycosidic linkages after methylation using iodomethane/sodium hydroxide, high-performance gel permeation chromatography for homogeneity and molecular weight, and FT-IR spectrometric analysis to identify functional groups [6–8]. Determining the chemical properties of EPS is crucial for their potential application in industries such as food, pharmaceuticals, and bioplastics. This study contributes to the growing knowledge of sustainable polysaccharide production using yeast, particularly through the utilization of *R. toruloides* as a microbial source. In conclusion, this study underscores the potential of *R. toruloides* as a promising candidate for microbial polysaccharide production. The optimization of cultivation conditions and EPS characterization pave the way for exploring yeast-based polysaccharides in various industrial applications.

Citation: Hamacek, H.S.D.R.; Lahtvee, P.-J.; Kumar, R. Production and Characterization of Polysaccharides from *Rhodotorula toruloides* . *Proceedings* **2023**, *92*, 15. https://doi.org/10.3390/proceedings2023092015

Published: 21 November 2023

Author Contributions: Conceptualization, H.S.D.R.H. and R.K.; methodology, H.S.D.R.H.; software, H.S.D.R.H.; validation, H.S.D.R.H.; formal analysis, H.S.D.R.H.; investigation, H.S.D.R.H. and R.K.; resources, H.S.D.R.H.; data curation, H.S.D.R.H.; writing—original draft preparation, H.S.D.R.H.; writing—review and editing, H.S.D.R.H. and R.K.; visualization, H.S.D.R.H.; supervision, R.K.; project administration, R.K.; funding acquisition, P.-J.L. and R.K. All authors have read and agreed to the published version of the manuscript.

Funding: This research was funded by the Estonian Research Council, team grant number PRG1101 and the APC was funded by PRG1101.

Institutional Review Board Statement: Not applicable.

Informed Consent Statement: Not applicable.

Data Availability Statement: Data sharing is not applicable to this article.

Conflicts of Interest: Henrique Sepulveda Del Rio Hamacek and Rahul Kumar declare no conflicts of interest. Petri-Jaan Lahtvee has a financial interest in Äio, a biotechnology company that has no relation to or influence on the present research. The funder, Estonian Research Council, had no role in the design of the study; in the collection, analyses, or interpretation of data; in the writing of the manuscript; or in the decision to publish the results.

References

1. Saadat, Y.R.; Khosroushahi, A.Y. Yeast exopolysaccharides and their physiological functions. *Folia Microbiol.* **2021**, *66*, 171–182. [CrossRef]
2. Buck, J.W.; Andrews, J.H. Attachment of the yeast Rhodosporidium toruloides is mediated by adhesives localized at sites of bud cell development. *Appl. Environ. Microbiol.* **1999**, *65*, 465–471. [CrossRef]
3. Breierová, E.; Kocková-Kratochvílová, A. Cryoprotective effects of yeast extracellular polysaccharides and glycoproteins. *Cryobiology* **1992**, *29*, 385–390. [CrossRef]
4. Pavlova, K.; Panchev, I.; Krachanova, M.; Gocheva, M. Production of an exopolysaccharide by Antarctic yeast. *Folia Microbiol.* **2009**, *54*, 343–348. [CrossRef]
5. Illarionov, A.; Lahtvee, P.-J.; Kumar, R. Potassium and Sodium Salt Stress Characterization in the Yeasts *Saccharomyces cerevisiae*, *Kluyveromyces marxianus*, and *Rhodotorula toruloides*. *Appl. Environ. Microbiol.* **2021**, *87*, e03100-20. [CrossRef]
6. Van Dyk, J.S.; Kee, N.L.A.; Pletschke, B.I. Extracellular polysaccharide production in Bacillus licheniformis SVD1 and its immunomodulatory effect. *BioResources* **2012**, *7*, 4976–4993. [CrossRef]
7. Li, H.; Huang, L.; Zhang, Y.; Yan, Y. Production, Characterization and Immunomodulatory Activity of an Extracellular Polysaccharide from Rhodotorula mucilaginosa YL-1 Isolated from Sea Salt Field. *Mar. Drugs* **2020**, *18*, 595. [CrossRef]
8. Hamidi, M.; Gholipour, A.R.; Delattre, C.; Sesdighi, F.; Seveiri, R.M.; Pasdaram, A.; Kheirandish, S.; Pierre, G.; Kozani, P.S.; Kozani, P.S.; et al. Production, characterization and biological activities of exopolysaccharides from a new cold-adapted yeast: *Rhodotorula mucilaginosa* sp. GUMS16. *Int. J. Biol. Macromol.* **2020**, *151*, 268–277. [CrossRef]

proceedings

MDPI

Abstract

Elemental Composition and Isotope Ratio in Pine Needles: The Impact of Arginine Phosphate-Containing Fertilizer Application in Pine-Planting Sites †

Maris Bertins [1,*], Jana Svinska [1], Sindija Zigure [2], Lauma Busa [1], Austra Zusevica [2], Karlis Dumins [2], Viktorija Vendina [2], Toms Arturs Stals [2], Linda Ansone-Bertina [3], Dagnija Lazdina [2], Maris Klavins [3] and Arturs Viksna [1]

1 Faculty of Chemistry, University of Latvia, Jelgavas Street 1, LV-1004 Riga, Latvia
2 Latvian State Forest Research Institute 'Silava', 111 Rigas Street, LV-2169 Salaspils, Latvia
3 Faculty of Geography and Earth Sciences, University of Latvia, Jelgavas Street 1, LV-1004 Riga, Latvia
* Correspondence: maris.bertins@lu.lv; Tel.: +371-29869037

† Presented at the International Conference EcoBalt 2023 "Chemicals & Environment", Tallinn, Estonia, 9–11 October 2023.

Keywords: pine needles; arginine phosphate; isotope ratio; chemometric analysis

Forests in Latvia are crucial, covering over half of the country's territory and expanding continuously through afforestation and natural growth. However, like forests globally, they face challenges requiring attention. Climate and biodiversity changes call for sustainable forest management practices different from those in the past to ensure the long-term health, resilience, and ecological value of Latvia's forests. This study investigates the impact of an arginine phosphate-containing fertilizer on nitrogen uptake, carbon content, and elemental concentrations in pine needles across different forest types in Latvia. By examining the effects of this fertilizer in the context of Latvia's specific forest-related issues, the research aims to contribute valuable insights into nutrient dynamics and concurrence in the first years after planting. The study encompasses three distinct forest types: Vacciniosa, Aegopodiosa, and Myrtillosa. Soil treatment was implemented during the planting of the seedlings, followed by the analysis of pine needle samples. Isotope ratio mass spectrometry and inductively coupled plasma mass spectrometry were employed to determine the nitrogen and carbon mass fraction, the nitrogen isotope ratio, and elemental concentrations. Chemometric analysis facilitated data evaluation. The findings reveal diverse patterns in nitrogen uptake and isotope ratio changes among the forest types. Aegopodiosa and Myrtillosa forests exhibited increased nitrogen mass fraction and decreased δ^{15}N values in pine needles, indicating arginine phosphate as the primary nitrogen source. Conversely, Vacciniosa forests displayed elevated δ^{15}N values in control samples, suggesting alternative nitrogen uptake due to low soil nitrogen content. All samples exhibited a significant increase in carbon content and a decrease in δ^{13}C values associated with transplantation and environmental shifts. Aegopodiosa forests demonstrated the least variation in δ^{13}C values, indicating a more consistent response during transplantation. Chemometric analysis highlighted correlations between elemental concentrations, seedling age, and forest types [1]. This study highlights the importance of considering forest type and environmental conditions when assessing fertilizer efficacy. It provides insights into the varying effects on nitrogen uptake and carbon content in pine needles across different forest types in Latvia, contributing to our understanding of nutrient dynamics in forest ecosystems and guiding sustainable forest management practices.

Citation: Bertins, M.; Svinska, J.; Zigure, S.; Busa, L.; Zusevica, A.; Dumins, K.; Vendina, V.; Stals, T.A.; Ansone-Bertina, L.; Lazdina, D.; et al. Elemental Composition and Isotope Ratio in Pine Needles: The Impact of Arginine Phosphate-Containing Fertilizer Application in Pine-Planting Sites. *Proceedings* **2023**, *92*, 50. https://doi.org/10.3390/proceedings2023092050

Published: 24 November 2023

Author Contributions: Conceptualization, M.B. and D.L.; methodology, A.V., D.L. and M.K.; software, M.B. and L.A.-B.; validation, D.L., M.K. and A.V.; formal analysis, M.B., J.S., L.B., K.D., A.Z,

V.V., T.A.S. and S.Z.; data curation M.B., J.S., L.B., K.D., A.Z., V.V., T.A.S. and S.Z.; resources, A.V., D.L. and M.K.; writing—original draft preparation, M.B., J.S. and L.A.-B.; writing—review and editing, M.B., L.A.-B., D.L., M.K. and A.V.; visualization, M.B., J.S. and L.A.-B.; supervision, D.L., M.K. and A.V. All authors have read and agreed to the published version of the manuscript.

Funding: Research was funded by the Project "Strengthening the Doctoral Capacity of the University of Latvia within the framework of the new doctoral model" with project identification No.8.2.2.0/20/I/006, LU registration No. ESS2021/434, co-financed by the European Social Fund. Trees were measured to determine the growth rate and potential of accumulation of elements in cooperation with the Latvian State Forest initiated research "Working methods and technologies for restoration, planting, care and protection of forest stands" No. 5-5.9.1_007O_101_21_77).

Institutional Review Board Statement: Not applicable.

Informed Consent Statement: Not applicable.

Data Availability Statement: The data presented in this study are available upon request from the corresponding author.

Acknowledgments: The authors would like to express their gratitude to all individuals who provided assistance and support throughout this research project. Their contributions and collaboration were invaluable in the successful completion of this study.

Conflicts of Interest: The authors declare no conflict of interest.

Reference

1. Bērtiņš, M.; Buša, L.; Lazdina, D.; Dumins, K.; Zake, S.; Kļaviņš, M.; Vīksna, A. Impact of Arginine Containing Fertilizer on Nitrogen Isotope Ratio and Elemental Content in Young Conifer Stands. *Key Eng. Mater.* **2022**, *933*, 185–192. [CrossRef]

proceedings

Abstract

Antioxidative and Anti-*Borrelia* Effects of *Plantago* Species †

Pille-Riin Laanet [1,*], Regina Drošnova [1], Piret Saar-Reismaa [1,2], Olga Bragina [1,3] and Merike Vaher [1]

1 Department of Chemistry and Biotechnology, Tallinn University of Technology, Akadeemia Tee 15, 12618 Tallinn, Estonia
2 HAN BioCentre, Laan van Scheut 2, 6525 EM Nijmegen, The Netherlands
3 National Institute for Health Development, Hiiu 42, 11619 Tallinn, Estonia
* Correspondence: pille-riin.laanet@taltech.ee

† Presented at the International Conference EcoBalt 2023 "Chemicals & Environment", Tallinn, Estonia, 9–11 October 2023.

Keywords: Lyme disease; *Borrelia burgdorferi*; phytochemicals; antioxidants

Borrelia burgdorferi sensu lato bacteria are the causative agent of Lyme disease, Europe's most common vector-borne disease. In Estonia, the number of ticks carrying pathogenic bacteria and the case numbers of the illness are rapidly rising [1]. The infection can affect multiple organ systems and withstand several rounds of antibiotic treatment [2]. Therefore, novel treatment options are needed to combat the persister forms of the bacteria responsible for the chronic illness [3]. The screening of natural resources has shown promise in helping discover lead compounds with distinct anti-*Borrelia* activity for future therapeutic approaches. The antioxidative and antibacterial properties of several plants found in Estonia have been demonstrated by our group. This presentation discusses the chemical characterisation and anti-*Borrelia* activity determination of *Plantago major* and *Plantago lanceolata*. The plants' main groups of bioactive compounds were quantified by colorimetric tests, total polyphenols by the Folin–Ciocalteu, total flavonoids by the $AlCl_3$, and total iridoids by the Trim–Hill method. The results show that dried aerial parts of *P. major* and *P. lanceolata* contain up to 32.7 and 47.1 mg/g gallic acid equivalents of phenolic compounds, up to 10.0 and 14.4 mg/g quercetin equivalents of flavonoids, and up to 11.4 and 23.4 mg/g asperuloside equivalents of iridoids, respectively. The extracts were chemically characterised using HPLC–DAD–MS/MS. The antioxidative activity of all extracts was evaluated using the $ORAC_{FL}$ method, and found to be up to 12.3 or 14.6 mg/g Trolox equivalents for *P. major* and *P. lanceolata*, respectively. The anti-*Borrelia* activity of the plant extracts was tested on the latent bacterial forms using the SYBR Green I and Propidium Iodide assay. The residual viability of *B. burgdorferi* bacteria after incubation with the plant extracts was as low as 18.7% for *P. major* species, and 23.6% for *P. lanceolata* species. Therefore, as our results demonstrate that both *P. major* and *P. lanceolata* contain considerable amounts of phytochemicals with antioxidant properties and show significant anti-*Borrelia* effects on the latent forms of *B. burgdorferi*, these plants should be considered for further therapeutic research.

Citation: Laanet, P.-R.; Drošnova, R.; Saar-Reismaa, P.; Bragina, O.; Vaher, M. Antioxidative and Anti-*Borrelia* Effects of *Plantago* Species. *Proceedings* **2023**, *92*, 6. https://doi.org/10.3390/proceedings2023092006

Academic Editors: Monika Mortimer, Anne Kahru, Ivo Leito, Riin Rebane and Villem Aruoja

Published: 21 November 2023

Author Contributions: Conceptualization: M.V.; methodology: P.S.-R. and M.V.; formal analysis: R.D., P.-R.L., P.S.-R. and O.B.; investigation: P.-R.L., P.S.-R. and M.V.; data curation: P.-R.L. and P.S.-R.; writing—original draft preparation: P.-R.L.; writing—review and editing: P.-R.L. and M.V.; visualization: P.-R.L.; supervision: P.S.-R. and M.V.; project administration: M.V.; funding acquisition: M.V. All authors have read and agreed to the published version of the manuscript.

Funding: This research was supported by the Estonian Center of Analytical Chemistry (ECAC) funded by the Estonian Research Council (TT4) and the R&D project SS220044 "Evaluation of antioxidant and antibacterial activity of plant extracts" funded by Tallinn University of Technology.

Institutional Review Board Statement: Not applicable.

Informed Consent Statement: Not applicable.

Data Availability Statement: The data presented in this study are available upon request from the corresponding author.

Conflicts of Interest: The authors declare no conflict of interest. The funders had no role in the design of the study; in the collection, analyses, or interpretation of data; in the writing of the manuscript; or in the decision to publish the results.

References

1. Geller, J.; Vikentjeva, M.; Bragina, O.; Jakovleva, J. Report: "Send Tick by Post!" Campaign of 2020 and Pathogens in Estonian Ticks. Available online: https://www.tai.ee/sites/default/files/2022-07/Puugiraport_2022.pdf (accessed on 26 June 2023).
2. Singh, S.K.; Girschick, H.J. Lyme borreliosis: From infection to autoimmunity. *Clin. Microbiol. Infect.* **2004**, *10*, 598–614. [CrossRef] [PubMed]
3. Sharma, B.; Brown, A.V.; Matluck, N.E.; Hu, L.T.; Lewis, K. Borrelia burgdorferi, the Causative Agent of Lyme Disease, Forms Drug-Tolerant Persister Cells. *Antimicrob. Agents Chemother.* **2015**, *59*, 4616–4624. [CrossRef] [PubMed]

proceedings

Abstract

The Study of the Uptake of Chromium, Zinc, Cadmium and Lead from Spiked Nutrient Solution in Tomato Plants †

Katarina Marković [1,*], Radmila Milačič [1,2], Ester Heath [1,2], Nina Kacjan Maršić [3], Marina Pintar [3] and Janez Ščančar [1,2]

1 Department of Environmental Sciences, Jožef Stefan Institute, 1000 Ljubljana, Slovenia; radmila.milacic@ijs.si (R.M.); ester.heath@ijs.si (E.H.); janez.scancar@ijs.si (J.Š.)
2 Jožef Stefan International Postgraduate School, 1000 Ljubljana, Slovenia
3 Biotechnical Faculty, Department of Agronomy, University of Ljubljana, Jamnikarjeva 101, 1000 Ljubljana, Slovenia; nina.kacjan.marsic@bf.uni-lj.si (N.K.M.); marina.pintar@bf.uni-lj.si (M.P.)
* Correspondence: katarina.markovic@ijs.si
† Presented at the International Conference EcoBalt 2023 "Chemicals & Environment", Tallinn, Estonia, 9–11 October 2023.

Keywords: wastewater; irrigation; Cr; Zn; Cd; Pb; food safety; ICP-MS

Citation: Marković, K.; Milačič, R.; Heath, E.; Maršić, N.K.; Pintar, M.; Ščančar, J. The Study of the Uptake of Chromium, Zinc, Cadmium and Lead from Spiked Nutrient Solution in Tomato Plants. *Proceedings* **2023**, *92*, 69. https://doi.org/10.3390/proceedings2023092069

Academic Editors: Monika Mortimer, Anne Kahru, Ivo Leito, Riin Rebane and Villem Aruoja

Published: 1 December 2023

The reuse of treated wastewater for irrigation can preserve freshwater resources. However, treated wastewater carries health risks due to residual contaminants. For safe food production, it is essential to understand the uptake of potential contaminants and their distribution in edible parts. Chromium (Cr), zinc (Zn), cadmium (Cd) and lead (Pb) are commonly found in municipal wastewaters. In the present study, the uptake of Cr, Zn, Cd and Pb was studied. Tomatoes were cultivated in Hoagland nutrient solution (pH 7) prepared using potable water and nutrients [1]. Treated tomatoes were grown in the Hoagland solution spiked with naturally abundant elements ^{52}Cr (100 ng/mL), ^{66}Zn (100 ng/mL), ^{111}Cd (50 ng/mL) and ^{208}Pb (100 ng/mL) and their enriched stable isotopes ^{53}Cr, ^{70}Zn, ^{106}Cd and ^{204}Pb at the same concentration levels. The use of enriched isotopes enabled distinguishing between the concentration of an individual element, which is naturally present in Hoagland's solution and is uptaken by plant during growth, and the concentration of an individual element uptaken by plant as a result of its addition in the nutrient solution. Tomatoes were raised 5 weeks in 40 L pots. Nutrient solution or spiked nutrient solution were weekly replenished to volume. Plants were harvested 5 weeks after exposure and divided into roots, stems, leaves and fruits. Samples were decomposed using microwave digestion, and element concentrations were determined by ICP-MS. The results showed that the roots exhibited the highest accumulation of Cr, Zn, Cd and Pb, while low concentrations of Cr and Pb were determined in fruits. In the fruits, the highest accumulation was observed for the essential element Zn. In compliance with Commission Regulation, the concentration of Cd in fruits exceeded four times the permissible levels set for fruiting-like vegetables. This study contributes to the understanding of pathways of accumulation of essential and toxic elements in plants and to food safety.

Author Contributions: Conceptualization, J.Š., R.M., E.H., N.K.M. and M.P.; methodology, R.M., J.Š., E.H., N.K.M. and M.P.; formal analysis, K.M.; investigation, R.M., J.Š. and K.M.; resources, E.H.; data curation, R.M., J.Š. and K.M.; writing—original draft preparation, K.M., R.M. and J.Š.; writing—review and editing: J.Š., R.M. and E.H. and visualization, K.M.; supervision, J.Š. and R.M.; project administration, E.H. All authors have read and agreed to the published version of the manuscript.

Funding: This research was funded by Slovene Research Agency, namely Program Group P1-0143 and Project L7-4422.

Institutional Review Board Statement: Not applicable.

Informed Consent Statement: Not applicable.

Data Availability Statement: The data presented in this study are available on request from the corresponding author.

Conflicts of Interest: The authors declare no conflict of interest.

Reference

1. Ahmed, D.A.E.-A.; Slima, D.F.; Al-Yasi, H.M.; Hassan, L.M.; Galal, T.M. Risk assessment of trace metals in *Solanum lycopersicum* L. (tomato) grown under wastewater irrigation conditions. *Environ. Sci. Pollut. Res.* **2023**, *30*, 42255–42266. [CrossRef]

proceedings

Abstract

Burning of Fountain Candles Indoor—A Moment of Joy versus Indoor Air Quality Concerns †

Agnese Araja [1,*], Maja Remskar [2,3], Maris Bertins [1], Laima Plaveniece [1] and Arturs Viksna [1]

1 Faculty of Chemistry, University of Latvia, Jelgavas Str. 1, LV-1004 Riga, Latvia; maris.bertins@lu.lv (M.B.); laima.plaveniece@gmail.com (L.P.); arturs.viksna@lu.lv (A.V.)
2 Solid State Physics Department, Jožef Stefan Institute, Jamova Str. 39, 1000 Ljubljana, Slovenia; maja.remskar@ijs.si
3 Faculty of Mathematics and Physics, University of Ljubljana, Jadranska Str. 19, 1000 Ljubljana, Slovenia
* Correspondence: agnese.araja@lu.lv
† Presented at the International Conference EcoBalt 2023 "Chemicals & Environment", Tallinn, Estonia, 9–11 October 2023.

Keywords: particulate matter; number concentration; fountain candles; indoor air; Grimm spectrometer

Citation: Araja, A.; Remskar, M.; Bertins, M.; Plaveniece, L.; Viksna, A. Burning of Fountain Candles Indoor—A Moment of Joy versus Indoor Air Quality Concerns. *Proceedings* **2023**, *92*, 38. https://doi.org/10.3390/proceedings2023092038

Academic Editors: Monika Mortimer, Anne Kahru, Ivo Leito, Riin Rebane and Villem Aruoja

Published: 23 November 2023

Burning candles, magic candles, incense, and pyrotechnics such as fountain candles creates a sense of joy and celebration in various life and social events. Unfortunately, most people never even thought about the smoke, particulate matter, or the pollution that occurs and remains in indoor air after such activities. A lot of studies have raised the issue of whether the use of such products indoors can worsen indoor air quality [1,2]. Contrary to that, the exposure to fine ambient particulate matter has been associated with cardiovascular and respiratory diseases, and its relevance to particulate matter from different candle burning remains unexplored [3].

The objective of the current study was to characterize the number concentration and mass concentration of particulate matter originated indoors when burning fountain candles, which are commercially available in Latvia. Therefore, simulating studies of burning fountain candles were performed in a close laboratory for quantitative analysis of obtained particulate matter. For the mass concentration of PM_{10}, $PM_{2.5}$, PM_1, and number concentration of ultra-fine and fine particulate matter with diameters from 0.265 up to 2.750 μm measuring, the spectrometer GRIMM EDM—365 (Grimm Aerosol Technik, Ainring, Germany) was used. In order to determine possible exposure of emitted particulate matter, a detecting device was placed 1 m from the fountain candle. In another experiment, a detecting device was placed 4 m from the fountain candle to characterize particulate matter distribution dynamics (see Figure 1).

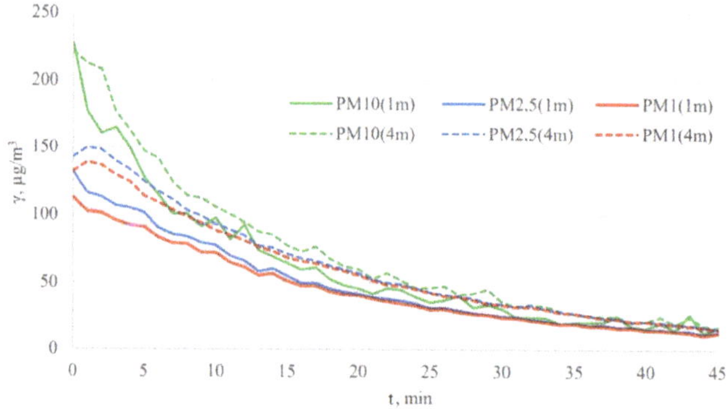

Figure 1. Mass concentrations of PM_{10}, $PM_{2.5}$, and PM_1 emitted from fountain candle burning depending on the distance of the detecting devise.

Model experiments showed that most of the particulate matter released indoors after fountain candles burning are in size $\leq 0.265 \mu m$, which are the most harmful to human health. The mass and number concentration of particles depend on the distance at which the emission source is located, e.g., ultra fine particles move further in the room and their concentrations are higher at 4 m compared to 1 m.

Author Contributions: Conceptualization, A.A. and M.R.; methodology, A.A. and M.R.; software, L.P.; validation, M.B. and L.P.; formal analysis, L.P.; investigation, A.A.; resources, A.V.; data curation, L.P.; writing—original draft preparation, A.A.; writing—review and editing, M.R.; visualization, L.P.; supervision, A.A.; project administration, A.V. All authors have read and agreed to the published version of the manuscript.

Funding: This research received no external funding.

Institutional Review Board Statement: Not applicable.

Informed Consent Statement: Not applicable.

Data Availability Statement: Data are unavailable.

Conflicts of Interest: The authors declare no conflict of interest.

References

1. Derudi, M.; Gelosa, S.; Sliepcevich, A.; Cattaneo, A.; Cavallo, D.; Rota, D.; Nano, G. Emission of air pollutants from burning candles with different composition in indoor environments. *Environ. Sci. Pollut. Res.* **2014**, *21*, 4320–4330. [CrossRef] [PubMed]
2. Niu, X.; Jones, T.; BéruBé, K.; Chuang, H.; Sune, J.; Ho, H.F. The oxidative capacity of indoor source combustion derived particulate matter and resulting respiratory toxicity. *Sci. Total Environ.* **2021**, *767*, 144391. [CrossRef] [PubMed]
3. Loft, S.; Andersen, Z.A.; Jørgensen, J.T.; Kristiansen, A.D.; Dam, J.K.; Cramer, J.; Rudi, G.J.; Westendorp, R.G.J.; Lund, R.; Lim, J.H. Use of candles and risk of cardiovascular and respiratory events in a Danish cohort study. *Indoor Air* **2022**, *32*, e13086. [CrossRef] [PubMed]

proceedings

Abstract

The Effects of Pesticides on the Bioenergetics of Intestinal Cells †

Giuseppe Leonardo Auditano [1,2], Karolin Kullison [2], Tuuli Kaambre [1] and Kersti Tepp [1,*]

1 Laboratory of Chemical Biology, National Institute of Chemical Physics and Biophysics, Akadeemia Tee 23, 12618 Tallinn, Estonia; giuseppeleonardoauditano@outlook.it (G.L.A.); tuuli.kaambre@kbfi.ee (T.K.)
2 Department of Chemistry and Biotechnology, School of Science, Tallinn University of Technology, Ehitajate Tee 5, 12618 Tallinn, Estonia; karolin.kullison@gmail.com
* Correspondence: kersti.tepp@kbfi.ee; Tel.: +372-56154632

† Presented at the International Conference EcoBalt 2023 "Chemicals & Environment", Tallinn, Estonia, 9–11 October 2023.

Keywords: mitochondria; pesticides; high-resolution respirometry; oxidative phosphorylation

Citation: Auditano, G.L.; Kullison, K.; Kaambre, T.; Tepp, K. The Effects of Pesticides on the Bioenergetics of Intestinal Cells. *Proceedings* **2023**, *92*, 66. https://doi.org/10.3390/proceedings2023092066

Academic Editors: Monika Mortimer, Anne Kahru, Ivo Leito, Riin Rebane and Villem Aruoja

Published: 30 November 2023

Colorectal cancer is considered the second most common type of cancer [1]. It is assumed for this type of cancer that environmental factors are more important than hereditary factors [2]. Pesticides present in our food can alter the metabolism of our intestinal cells, and this might be related to the potential impact on overall cellular metabolism. The aim of this project was to investigate whether the pesticides Glyphosate, Glyphosate-based Roundup, Boscalid, and NeemAzal alter the energy metabolism of human intestinal cells. The study analyses if prolonged exposure and different growth environments increase the sensitivity of Caco-2 cells to pesticides. The rate of oxygen consumption based on electron flow through individual respiratory chain complexes and the overall oxygen consumption rate of the respiratory chain were analyzed using the method of high-resolution respirometry. The results demonstrated that lower concentrations of pesticides, which do not affected cells in the short term, significantly decreased cell viability with prolonged use. The experiments also showed that, in a plasma-like medium, similar to physiological conditions, the toxic effect of pesticides is higher or equivalent to that observed in cells grown in a regular medium. Analysis of mitochondrial oxidative phosphorylation revealed a significant decrease in the oxygen consumption rate through the electron transport chain at concentrations reducing cell viability by 20% for all pesticides. At lower pesticide concentrations reducing viability by up to 10%, the effect was detectable only for Boscalid and Roundup. The results of the study confirm that commercially available pesticide Roundup, along with its accompanying additives, exhibits stronger toxic effects than the declared active ingredient Glyphosate alone. The results of the study indicate that low pesticide concentrations, which have no immediate impact, may exert toxic effects over a longer period, and this influence should be studied in a plasma-like medium.

Author Contributions: Conceptualization, K.T. and T.K.; methodology, K.T.; formal analysis, K.T.; investigation, K.T.; G.L.A. and K.K.; writing—original draft preparation, G.L.A. and K.T.; writing—review and editing, T.K.; visualization, G.L.A. and K.T.; supervision, K.T. and T.K.; funding acquisition, K.T. All authors have read and agreed to the published version of the manuscript.

Funding: This research was funded by National Institute of Chemical Physics and Biophysics project Arengufond_KT.

Institutional Review Board Statement: Not applicable.

Informed Consent Statement: Not applicable.

Data Availability Statement: An article based on these data is in the process of submission.

Conflicts of Interest: The authors declare no conflict of interest.

References

1. Siegel, R.L.; Wagle, N.S.; Cercek, A.; Smith, R.A.; Jemal, A. Colorectal cancer statistics. *CA Cancer J. Clin.* **2023**, *73*, 233–254. [CrossRef] [PubMed]
2. Tian, J.; Afebu, K.O.; Bickerdike, A.; Liu, Y.; Prasad, S.; Nelson, B.J. Fundamentals of Bowel Cancer for Biomedical Engineers. *Ann. Biomed. Eng.* **2023**, *51*, 679–701. [CrossRef] [PubMed]

proceedings

MDPI

Abstract

Variation of Carbon- and Nitrogen-Stable Isotope Ratios in Conventionally and Organically Fertilized Cereals at Different Growth Stages †

Lauma Busa [1,*], Gunita Celma [1], Linda Legzdina [2], Arturs Viksna [1] and Indra Locmele [2]

1 Faculty of Chemistry, University of Latvia, 1004 Riga, Latvia; gunita.celma@lu.lv (G.C.); arturs.viksna@lu.lv (A.V.)
2 Institute of Agricultural Resources and Economics, Priekuli Research Centre, 4126 Priekuli, Latvia; linda.legzdina@arei.lv (L.L.)
* Correspondence: lauma.busa@lu.lv; Tel.: +371-26244773
† Presented at the International Conference EcoBalt 2023 "Chemicals & Environment", Tallinn, Estonia, 9–11 October 2023.

Keywords: stable isotope ratios; barley; triticale; organic farming; conventional farming; cereal grains; cereal growth stages

Over recent decades, the cereal grain market has experienced notable changes. Due to the demand for healthier and more sustainable food options, significant growth in the production of organically grown cereals has been observed in the EU and Latvia [1]. Climate change and its impact on agricultural productivity has created challenges to cereal grain production, leading to a focus on resilient crop varieties and sustainable farming practices. Research on stable carbon and nitrogen isotope ratio changes in barley and triticale at different growth stages provides valuable insights into the metabolic processes and nutrient uptake patterns of these crops. This research contributes to the improvement of sustainable agricultural practices by allowing the optimization of fertilization strategies and the development of more efficient crop management techniques. For this study, barley and triticale samples at the stages of tillering, jointing, booting and maturity stages from conventionally and organically fertilized sample plots were collected at the Institute of Agricultural Resources and Economics, Priekuli Research Centre. Roots, leaves and grains at maturity stage of the collected crop samples were analyzed using a stable isotope ratio mass spectrometer (Nu Horizon, Nu Instruments, Wrexham, UK). $\delta^{13}C$ and $\delta^{15}N$ values, and total carbon and nitrogen content were determined. The results showed a decrease in $\delta^{15}N$ values and total nitrogen content in both barley and triticale roots and leaves during the growth of the analyzed crop samples. No significant changes in $\delta^{13}C$ values and total carbon content were observed. Differences in total nitrogen content and nitrogen-stable isotope ratios between conventionally and organically fertilized crops were not definite both for barley and triticale samples. These findings highlight the dynamic nature of nitrogen uptake and utilization in barley and triticale crops during various growth stages and suggest that other factors beyond fertilizer type may influence nitrogen content and isotope ratios in these crops.

Citation: Busa, L.; Celma, G.; Legzdina, L.; Viksna, A.; Locmele, I. Variation of Carbon- and Nitrogen-Stable Isotope Ratios in Conventionally and Organically Fertilized Cereals at Different Growth Stages. *Proceedings* **2023**, *92*, 19. https://doi.org/10.3390/proceedings2023092019

Academic Editors: Monika Mortimer, Anne Kahru, Ivo Leito, Riin Rebane and Villem Aruoja

Published: 21 November 2023

Author Contributions: Conceptualization, L.B. and A.V.; methodology, A.V., L.B., L.L. and I.L.; software, L.B.; validation, L.B., A.V. and G.C.; formal analysis, L.B. and G.C.; investigation, L.B. and G.C.; resources, A.V., L.L. and I.L.; data curation, L.B. and G.C.; writing—original draft preparation, L.B. and G.C.; writing—review and editing, L.B., G.C. and A.V.; visualization, L.B.; supervision, A.V., L.L. and I.L. All authors have read and agreed to the published version of the manuscript.

Funding: This research was financially supported by Ministry of Agriculture of Latvia and Latvian Council of Science, grant number lzp-2018/1-0404, acronym FLPP-2018-1.

Institutional Review Board Statement: Not applicable.

Informed Consent Statement: Not applicable.

Data Availability Statement: The data presented in this study are available upon request from the corresponding author.

Acknowledgments: The authors would like to express their gratitude to all individuals who provided assistance and support throughout this research project. Their contributions and collaboration were invaluable in the successful completion of this study.

Conflicts of Interest: The authors declare no conflict of interest. The funders had no role in the design of the study; in the collection, analyses, or interpretation of data; in the writing of the manuscript; or in the decision to publish the results.

Reference

1. Busacca, E.; Gernert, M.; Moeskops, B.; Steu, A.; Willer, H. Organic in Europe: Recent Developments. In *The World Of Organic Agriculture. Statistics and Emerging Trends 2023*; Willer, H., Schlatter, B., Travnicek, J., Eds.; Research Institute of Organic Agriculture FiBL, Frick, and IFOAM—Organics International: Bonn, Germany, 2023; pp. 225–233. [CrossRef]

proceedings

Abstract

Analysis and Usage Perspective of Solid Digestate †

Eglė Didžiulytė * and Rasa Šlinkšienė

Department of Physical and Inorganic Chemistry, Kaunas University of Technology, LT-50254 Kaunas, Lithuania; rasa.slinksiene@ktu.lt
* Correspondence: egle.didziulyte@ktu.lt

† Presented at the International Conference EcoBalt 2023 "Chemicals & Environment", Tallinn, Estonia, 9–11 October 2023.

Keywords: digestate; organic farming; organic carbon

Extremely intensive and chemicalized agriculture severely damages the naturally formed eco-system [1]. This issue encourages concern about soil conservation and improvement of its quality, human health, and environmental protection; therefore, it is necessary to look for alternative methods of agriculture. One of these methods is organic farming, where mineral fertilizers are not used and nutrients needed for plants are provided using biological substances of organic origin. On the other hand, the accumulation of such biodegradable organic waste and its disposal in landfills causes various environmental, economic, and social problems. One such waste is digestate [2]. Most of the digestate (about 120 million tons) is agricultural digestate (a mixture of manure and plants). The rest is obtained by mechanical biological treatment from municipal solid waste, separated biological waste, sewage sludge, or the agriculture and food industries. In various EU countries, renewable energy policies and subsidies for the production of electricity, gas, and heat from biomass have improved economic conditions for the anaerobic digestion of biological waste or food waste, but in this process, digestate is also produced. However, it was also noted that a lot of detailed scientific research should be conducted concerning the risk factors, environmental effects, fertilizer treatment methods, and enrichment during long-term use [3–5]. Because the digestate is rich in organic matter, during this study, when three different digestates were used, the organic carbon content was first determined. The tests were carried out on dried digestates, which were of two types: whole and chopped. It was found that the content of organic carbon is higher in whole samples and ranges from 39.0 to 50.0%, and ranges from 34.8 to 45.2% in chopped. In order to obtain the maximum concentration of organic carbon from the used digestates, extracts were produced on the basis of water, potassium alkali, and different production conditions. The samples were kept for 3, 6, and 9 days at room temperature and for 3, 6, and 9 h at different temperatures, mixed with a mechanical stirrer, and in an ultrasonic bath. Also, the samples were heated at temperatures of 50 °C, 70 °C, and 90 °C. A higher concentration of organic carbon was found in potassium-based extracts, and the higher temperature had a positive effect on the concentration of organic carbon in the samples.

Citation: Didžiulytė, E.; Šlinkšienė, R. Analysis and Usage Perspective of Solid Digestate. *Proceedings* **2023**, *92*, 34. https://doi.org/10.3390/proceedings2023092034

Academic Editors: Monika Mortimer, Anne Kahru, Ivo Leito, Riin Rebane and Villem Aruoja

Published: 22 November 2023

Author Contributions: Conceptualization, R.Š.; methodology, E.D. and R.Š.; software, E.D.; resources, E.D.; data curation, R.Š. and E.D.; writing—original draft preparation, E.D.; writing—review and editing, R.Š. All authors have read and agreed to the published version of the manuscript.

Funding: This research received no external funding.

Institutional Review Board Statement: Not applicable.

Informed Consent Statement: Not applicable.

Data Availability Statement: Data sharing not applicable.

Conflicts of Interest: The authors declare no conflict of interest.

References

1. Parikh, S.J.; James, B.R. Soil: The Foundation of Agriculture. *Nat. Educ. Knowl.* **2012**, *3*, 2.
2. Paull, J. Attending the First Organic Agriculture Course: Rudolf Steiner's Agriculture Course at Koberwitz, 1924. *Eur. J. Soc. Sci.* **2011**, *21*, 64–70.
3. Kovačić, Đ.; Lončarić, Z.; Jović, J.; Samac, D.; Popović, B.; Tišma, M. Digestate Management and Processing Practices: A Review. *Appl. Sci.* **2022**, *12*, 9216. [CrossRef]
4. Cheffins, N.; Stainton, D. *Final Project Report for OMK006-06*; WRAP: Banbury, UK, 2015.
5. Baştabak, B.; Koçar, G. A review of the biogas digestate in agricultural framework. *J. Mater. Cycles Waste Manag.* **2020**, *22*, 1318–1327. [CrossRef]

proceedings

Abstract

Exploring Gut Microbiota Metabolism—New Chemical Biology Tools for Metabolomics Analysis †

Vladyslav Dovhalyuk [1,*], Amanpreet Kaur [1], Weifeng Lin [1], Ioanna Tsiara [1], Sydney Mwasambu [1] and Daniel Globisch [1,2,*]

1 Department of Chemistry-BMC, Science for Life Laboratory, Uppsala University, 751 23 Uppsala, Sweden; amanpreet.kaur@kemi.uu.se (A.K.); weifeng.lin@kemi.uu.se (W.L.); ioanna.tsiara@kemi.uu.se (I.T.); sydney.mwasambu@kemi.uu.se (S.M.)
2 Centre for Translational Microbiome Research (CTMR), Karolinska Institute, 171 77 Solna, Sweden
* Correspondence: vladyslav.dovhalyuk@kemi.uu.se (V.D.); daniel.globisch@kemi.uu.se (D.G.)
† Presented at the International Conference EcoBalt 2023 "Chemicals & Environment", Tallinn, Estonia, 9–11 October 2023.

Keywords: metabolomics; chemoselective probe; gut microbiota; LC-MS/MS; mass spectrometry

The impact of the gut microbiota on human physiology through their vast metabolic activities has surfaced as a remarkable scientific discovery over the past decade. It has provided new avenues for biomarker discovery, especially via the analysis of known and unknown metabolites using mass spectrometry-based metabolomics. In the face of limited tools in chemical biology for metabolomics, we have designed advanced methodologies utilizing ultra-performance liquid chromatography coupled with tandem mass spectrometry (UPLC-MS/MS) [1–4], with an intent to uncover unknown metabolites from human samples, specifically from pancreatic cancer patients.

Understanding the bioactive metabolites produced by the gut microbiome, especially thiol-containing metabolites, is key to the discovery of potential novel drug scaffolds and dietary or disease biomarkers. To overcome the challenge posed by the lack of specific tools for analyzing thiol-containing metabolites, we have synthesized a unique chemoselective probe coupled to magnetic beads. This facilitates easy extraction of metabolites and enhances the mass spectrometric sensitivity significantly. Application of this technique on fecal samples has unveiled previously unknown metabolites and boosted the detection limit for most metabolites [1–4].

In a novel approach, we have incorporated bicyclobutane into our methodology for the chemoselective and irreversible capturing of thiol metabolites. Applying this tool to human plasma, fecal samples, and bacterial cultures, we have identified the core bacterial thiol metabolome of 394 features and specific bacterial metabolites for each bacterium, including first-time detections in human plasma [1].

This novel chemical biology method surmounts analytical limitations and simplifies the investigation of bioactive thiol-containing metabolites [1]. It uncovers a host of previously unidentified metabolites from dietary, bacterial, and human origins and paves the way for comprehensive mass spectrometric investigations. This holds significant potential for the discovery of disease biomarkers and therapeutic interventions.

Citation: Dovhalyuk, V.; Kaur, A.; Lin, W.; Tsiara, I.; Mwasambu, S.; Globisch, D. Exploring Gut Microbiota Metabolism—New Chemical Biology Tools for Metabolomics Analysis. *Proceedings* **2023**, *92*, 64. https://doi.org/10.3390/proceedings2023092064

Academic Editors: Monika Mortimer, Anne Kahru, Ivo Leito, Riin Rebane and Villem Aruoja

Published: 29 November 2023

Author Contributions: Conceptualization, D.G.; methodology, V.D., A.K., W.L., I.T. and S.M.; validation, V.D., A.K. and W.L.; formal analysis, V.D., A.K., W.L., I.T. and S.M.; investigation, V.D., A.K. and W.L.; resources, D.G.; data curation, V.D., A.K., W.L., I.T. and S.M.; writing—original draft preparation, V.D.; writing—review and editing, V.D. and D.G.; visualization, V.D., A.K. and W.L.; supervision, D.G.; project administration, D.G.; funding acquisition, D.G. All authors have read and agreed to the published version of the manuscript.

Funding: We are grateful for funding by the Swedish Research Council (VR 2016-04423/VR 2020-04707), the Swedish Cancer Foundation (19 0347 Pj), and the Science for Life Laboratory (SLL 2016/5) to D.G.

Institutional Review Board Statement: Not applicable.

Informed Consent Statement: Not applicable.

Data Availability Statement: Not applicable.

Acknowledgments: This study made use of the NMR Uppsala infrastructure, which is funded by the Department of Chemistry–BMC and the Disciplinary Domain of Medicine and Pharmacy.

Conflicts of Interest: The authors declare no conflict of interest.

References

1. Kaur, A.; Lin, W.; Dovhalyuk, V.; Driutti, L.; Di Martino, M.L.; Vujasinovic, M.; Löhr, J.-M.; Sellin, M.E.; Globisch, D. Chemoselective bicyclobutane-based mass spectrometric detection of biological thiols uncovers human and bacterial metabolites. *Chem. Sci.* **2023**, *14*, 5291–5301. [CrossRef] [PubMed]
2. Lin, W.; Conway, L.P.; Vujasinovic, M.; Löhr, J.-M.; Globisch, D. Chemoselective and Highly Sensitive Quantification of Gut Microbiome and Human Metabolites. *Angew. Chem. Int. Ed.* **2021**, *60*, 23232–23240. [CrossRef] [PubMed]
3. Garg, N.; Conway, L.P.; Ballet, C.; Correia, M.S.; Olsson, F.K.; Vujasinovic, M.; Löhr, J.-M.; Globisch, D. Chemoselective Probe Containing a Unique Bioorthogonal Cleavage Site for Investigation of Gut Microbiota Metabolism. *Angew. Chem. Int. Ed.* **2018**, *57*, 13805–13809. [CrossRef] [PubMed]
4. Conway, L.P.; Garg, N.; Lin, W.; Vujasinovic, M.; Löhr, J.-M.; Globisch, D. Chemoselective probe for detailed analysis of ketones and aldehydes produced by gut microbiota in human samples. *Chem. Commun.* **2019**, *55*, 9080–9083. [CrossRef] [PubMed]

proceedings

Abstract

Extraction and Determination of Total Phenolic Contents, Flavonoid Contents, and Volatile Compounds in *Epilobium angustifolium* and *Cannabis sativa* Varieties †

Kristin Düüna, Piia Jõul * and Merike Vaher

Department of Chemistry and Biotechnology, Tallinn University of Technology, Akadeemia tee 15, 12618 Tallinn, Estonia; krduun@taltech.ee (K.D.); merike.vaher@taltech.ee (M.V.)
* Correspondence: piia.joul@taltech.ee

† Presented at the International Conference EcoBalt 2023 "Chemicals & Environment", Tallinn, Estonia, 9–11 October 2023.

Keywords: *Epilobium angustifolium*; phenolic compounds; flavonoids; colorimetry; plant extracts; volatile compounds; *Cannabis sativa*; solid-phase microextraction

Plants are a natural source of phytochemicals, many of which have favorable bioactive properties for human health and are therefore used in ethnomedicine for preventing and treating a variety of diseases [1]. *Epilobium angustifolium* (fireweed) is a popular medical plant that is known and used worldwide. Due to its abundance of secondary metabolites such as polyphenols, tannins, and terpenes, fireweed has antioxidant, anti-inflammatory, and anti-aging properties [2]. Cannabis sativa (industrial hemp) is mainly cultivated as a source of industrial fiber or seed oil. In addition to cannabinoids, the chemical composition of industrial hemp includes many other active compound groups, mostly tannins and polyphenols [3]. The biological activity of the herb is dependent on its geographical origin and the sample preparation procedure [4].

The aim of the present study was to quantitatively determine the total phenolic and flavonoid contents and characterize the volatile compositions of three different plants growing in Estonia: fireweed and the industrial hemp varieties Finola and Estica. The volatile compounds were extracted from dried samples using headspace SPME (solid-phase microextraction) and analyzed through use of GC-MS. Polyphenols and flavonoids were investigated in plant ethanol extracts through use of colorimetric tests. As a result of SPME-GC-MS, over 15 volatiles were quantified ($\geq$1%) in each sample. The volatile compound compositions were similar in all of the samples, and the volatiles with the highest contents detected were β-caryophyllene and humulene. Colorimetric tests showed high concentrations of polyphenols (up to 157.6 $\pm$ 8.1 mg GAE/g) and flavonoids (up to 11.7 $\pm$ 0.9 mg QE/g), whereas the fireweed extracts indicated higher levels. The obtained results show that the investigated plants, fireweed, Finola, and Estica, are valuable sources of phenolic and volatile phytochemicals.

Citation: Düüna, K.; Jõul, P.; Vaher, M. Extraction and Determination of Total Phenolic Contents, Flavonoid Contents, and Volatile Compounds in *Epilobium angustifolium* and *Cannabis sativa* Varieties. *Proceedings* **2023**, *92*, 21. https://doi.org/10.3390/proceedings2023092021

Academic Editors: Monika Mortimer, Anne Kahru, Ivo Leito, Riin Rebane and Villem Aruoja

Published: 21 November 2023

Author Contributions: Conceptualization, P.J. and M.V.; methodology, software, validation, formal analysis, and investigation, P.J. and K.D.; resources, M.V.; data curation, writing—original draft preparation, writing—review and editing, and visualization, P.J. and K.D.; supervision, P.J.; project administration and funding acquisition, M.V. All authors have read and agreed to the published version of the manuscript.

Funding: This research was supported by the Estonian Center of Analytical Chemistry (ECAC) funded by the Estonian Research Council (TT4) and the R&D project SS22004 "Evaluation of antioxidant and antibacterial activity of plant extracts" funded by Tallinn University of Technology.

Institutional Review Board Statement: Not applicable.

Informed Consent Statement: Not applicable.

Data Availability Statement: The data presented in this study are available on request from the corresponding author. The data are not publicly available due to the still on-going research.

Acknowledgments: Tammejuure Organic Farm and Kanepi Municipality Government are acknowledged for kindly providing the samples.

Conflicts of Interest: The authors declare no conflict of interest. The funders had no role in the design of the study; in the collection, analysis, or interpretation of the data; in the writing of the manuscript; or in the decision to publish the results.

References

1. Salmerón-Manzano, E.; Garrido-Cardenas, J.A.; Manzano-Agugliaro, F. Worldwide Research Trends on Medicinal Plants. *Int. J. Environ. Res. Public Health* **2020**, *17*, 3376. [CrossRef]
2. Nowak, A.; Zagórska-Dziok, M.; Ossowicz-Rupniewska, P.; Makuch, E.; Duchnik, W.; Kucharski, Ł.; Adamiak-Giera, U.; Prowans, P.; Czapla, N.; Bargiel, P.; et al. *Epilobium angustifolium* L. Extracts as Valuable Ingredients in Cosmetic and Dermatological Products. *Molecules* **2021**, *26*, 3456. [CrossRef]
3. Andre, C.M.; Hausman, J.F.; Guerriero, G. *Cannabis sativa*: The Plant of the Thousand and One Molecules. *Front. Plant Sci.* **2016**, *7*, 19. [CrossRef]
4. Kaškonienė, V.; Stankevičius, M.; Drevinskas, T.; Akuneca, I.; Kaškonas, P.; Bimbiraitė-Survilienė, K.; Maruška, A.; Ragažinskienė, O.; Kornyšova, O.; Briedis, V.; et al. Evaluation of phytochemical composition of fresh and dried raw material of introduced Chamerion angustifolium L. using chromatographic, spectrophotometric and chemometric techniques. *Phytochemistry* **2015**, *115*, 184–193. [CrossRef]

proceedings

MDPI

Abstract

The Interaction between Extracellular Polymeric Substances from Diatom *Cyclotella meneghiniana* and Citrate-Coated Silver Nanoparticles †

Rocco Gasco *, Isabelle A. Worms and Vera I. Slaveykova

Department F.-A. Forel for Environmental and Aquatic Sciences, Environmental Biogeochemistry and Ecotoxicology, Faculty of Sciences, Earth and Environment Sciences, University of Geneva, 66 Boulevard Carl-Vogt, CH-1211 Geneva, Switzerland; isabelle.worms@unige.ch (I.A.W.); vera.slaveykova@unige.ch (V.I.S.)

* Correspondence: rocco.gasco@unige.ch; Tel.: +41-22-379-03-06

† Presented at the International Conference EcoBalt 2023 "Chemicals & Environment", Tallinn, Estonia, 9–11 October 2023.

Keywords: silver nanoparticles; EPS; diatom

Opposite to the significant knowledge about the toxicity of AgNPs to phytoplankton species, rather limited knowledge is available about the role of phytoplankton secretions such as extracellular polymer substances (EPS) in NPs' fate [1]. This study, therefore, explores the interaction between 20 nm citrate-coated AgNPs (Cit-AgNPs) and the EPS derived from the diatom *Cyclotella meneghiniana*, and, in particular, their colloidal stability and transformations. To this end, a combination of different state-of-the-art techniques was employed to characterize the AgNPs-EPS interactions: Asymmetric Flow Field-Flow Fractionation (AF4) coupled with ICP-MS, Dynamic Light Scattering (DLS), Zeta Potential, and Surface Plasmon Resonance (SPR) absorbance spectroscopy. The changes in the size distribution, surface properties, and stability of Cit-AgNPs (4 mg/L) in the presence of various concentrations of EPS (130, 65, 32.5, and 13 mg C/L) were studied in both short-term (0–2 h) and long-term (72 h) experiments. The results showed that EPS stabilizes Cit-AgNPs, presumably through the formation of an ecocorona. The interaction occurred rapidly in the short term, leading to long-term stabilization, as revealed by the changes in the SPR-UV-Vis spectrum, characterized by the appearance of a typical shoulder associated with AgNPs' aggregation/agglomeration and alterations in hydrodynamic diameter. In the presence of EPS, the surface charge (zeta potential) of the Cit-AgNPs also shifted towards values similar to those of EPS alone, indicating an interaction. The EPS reduced the dissolution of Cit-AgNPs after 24 h. AF4-MD-ICP-MS provides relevant information with respect to DLS and UV-Vis spectroscopy, confirming the change in size of the Cit-AgNPs, the stabilization in the long term, and the interaction with the EPS. Taken together, the results of this study demonstrate that the EPS derived from the diatom *Cyclotella meneghiniana* modifies the surface properties and stability of 20 nm Cit-AgNPs. Therefore, such interactions have to be taken into consideration for predicting the fate and effects of Cit-AgNPs in the environment.

Citation: Gasco, R.; Worms, I.A.; Slaveykova, V.I. The Interaction between Extracellular Polymeric Substances from Diatom *Cyclotella meneghiniana* and Citrate-Coated Silver Nanoparticles. *Proceedings* **2023**, *92*, 48. https://doi.org/10.3390/proceedings2023092048

Academic Editors: Monika Mortimer, Anne Kahru, Ivo Leito, Riin Rebane and Villem Aruoja

Published: 24 November 2023

Author Contributions: Conceptualization, R.G. and V.I.S.; methodology, R.G., I.A.W. and V.I.S.; software, R.G., V.I.S.; validation, R.G. and V.I.S.; formal analysis, R.G., I.A.W.; investigation, R.G.; resources, V.I.S.; data curation, R.G., I.A.W. and V.I.S.; writing—original draft preparation, R.G. and V.I.S.; writing—review and editing, R.G. and V.I.S.; visualization, R.G. and V.I.S.; supervision, V.I.S.; project administration, V.I.S.; funding acquisition, V.I.S. All authors have read and agreed to the published version of the manuscript.

Funding: This research was funded by Swiss National Science Foundation No 204174.

Institutional Review Board Statement: Not applicable.

Informed Consent Statement: Not applicable.

Data Availability Statement: The data presented in this study are available upon request from the corresponding author.

Acknowledgments: The author wants to thank Arin Kantarciyan for the constructive discussion regarding the project.

Conflicts of Interest: The authors declare no conflict of interest.

Reference

1. Chetwynd, A.J.; Lynch, I. The Rise of the Nanomaterial Metabolite Corona, and Emergence of the Complete Corona. *Environ. Sci. Nano* **2020**, *7*, 1041–1060. [CrossRef]

proceedings

MDPI

Abstract

Revisiting Nanotoxicology Tests—Miniaturized Approaches of Nanotoxicity Tests in Daphnids †

Anne Leung [1], Emma Rowan [1], Dimitrios Kakavas [1], Konstantinos Panagiotidis [1], Keith D. Rochfort [2] and Konstantinos Grintzalis [1,*]

1 School of Biotechnology, Dublin City University, D09DX63 Dublin, Ireland; anne.leung@dci.ie (A.L.); emma.rowan@dcu.ie (E.R.); dimitrios.kakavas2@mail.dcu.ie (D.K.); konstantinos.panagiotidis@dcu.ie (K.P.)

2 School of Nursing, Psychotherapy and Community Health, Dublin City University, D09DX63 Dublin, Ireland; keith.rochfort@dcu.ie

* Correspondence: konstantinos.gkrintzalis@dcu.ie

† Presented at the International Conference EcoBalt 2023 "Chemicals & Environment", Tallinn, Estonia, 9–11 October 2023.

Abstract: The great increase in nanotechnology in the last 20 years has led to the alarming presence of nanomaterials in the environment as a new category of pollutants. Given the fact that legislation on nanomaterials is not concrete, the monitoring of their toxicity responses remains central. Focusing on tests in aquatic environments, daphnids are commonly employed as a bioindicator species for experiments with nanomaterials. However, until now, there has not been a unified and agreed approach to nanotoxicity testing, while research among different laboratories has been performed with significantly different setups, which may affect the reproducibility of the results. In this study, daphnids were exposed to silver nanoinks and the impact of surface to volume was assessed by comparing shallow vessels such as Petri dishes with deeper exposure vessels. Furthermore, in an attempt to assess whether the tests can be performed in smaller volumes, and thus in miniaturized versions, experiments compared larger and smaller volume setups. Finally, another parameter explored was the crowding of animals in exposure, and therefore their absolute number. Mortality was affected by both surface to volume and miniaturization, and significantly with crowding, supporting the implication of the number of animals in the tests. Further investigation with molecular and phenotypic endpoints confirmed these changes.

Keywords: molecular ecotoxicology; *Daphnia*; metabolomics; feeding; nanoink; nanotoxicology

Citation: Leung, A.; Rowan, E.; Kakavas, D.; Panagiotidis, K.; Rochfort, K.D.; Grintzalis, K. Revisiting Nanotoxicology Tests—Miniaturized Approaches of Nanotoxicity Tests in Daphnids. *Proceedings* **2023**, *92*, 52. https://doi.org/10.3390/proceedings2023092052

Academic Editors: Monika Mortimer, Anne Kahru, Ivo Leito, Riin Rebane and Villem Aruoja

Published: 27 November 2023

Author Contributions: Conceptualization, resources, writing—original draft preparation, writing—review and editing, supervision, funding acquisition, K.G. and K.D.R.; investigation, data acquisition, E.R., A.L., D.K. and K.P. All authors have read and agreed to the published version of the manuscript.

Funding: This research was funded by SCIENCE FOUNDATION IRELAND under grant number [18/SIRG/5563 Metabolomic approaches in mechanistic toxicology]. The IRISH RESEARCH COUNCIL supported Dimitrios Kakavas under grant number [GOIPG/2022/314 Mechanistic insight on the impact of nanomaterials and nanocoronas on freshwater and in vitro systems] and Dr. Konstantinos Panagiotidis under the grant number [GOIPD/2021/461 Nanoparticle metabolite coronas: A neglected feature with important contribution to toxicity].

Institutional Review Board Statement: Ethical review and approval were waived for this study, due to the fact that daphnids are regarded as "animals" in terms of being members of the kingdom Animalia, however, they are not "animals" as defined in regulation SI543 of 2012 on the protection of animals used for scientific purposes. Therefore, the study does not require authorization from the Health Products Regulatory Authority (HPRA), while is also in line with the aim of working under the 3Rs (reduce, refine, replacement) strategy, since daphnids are commonly used in ecology and ecotoxicology as replacements of more evolutionary advanced species (i.e., fishes), posing no ethical implications.

Informed Consent Statement: Not applicable.

Data Availability Statement: Not applicable.

Acknowledgments: The authors acknowledge the support from the microscopy facility of Dublin City University.

Conflicts of Interest: The authors declare no conflict of interest. The funders had no role in the design of the study; in the collection, analyses, or interpretation of data; in the writing of the manuscript; or in the decision to publish the results.

proceedings

MDPI

Abstract

Towards the Development of Sustainable Antimicrobial Surface Coatings †

Angela Ivask [1,*], Merilin Rosenberg [1] and Vambola Kisand [2]

[1] Institute of Molecular and Cell Biology, University of Tartu, 51010 Tartu, Estonia; merilin.rosenberg@ut.ee
[2] Institute of Physics, University of Tartu, 50411 Tartu, Estonia; vambola.kisand@ut.ee
* Correspondence: angela.ivask@ut.ee; Tel.: +372-53982998

† Presented at the International Conference EcoBalt 2023 "Chemicals & Environment", Tallinn, Estonia, 9–11 October 2023.

Keywords: antimicrobial materials; surface coatings; antimicrobial resistance; safety; sustainability

With the increasing trend of hard-to-treat microbial infections, including multiresistant nosocomial infections, food-related outbreaks, and the rapid spread of potentially infectious microbes in the common spaces of densely populated areas, awareness of the importance of proper systemic hygiene practices has increased. One of the main routes of potential pathogen transmission to vulnerable hosts is via contaminated surfaces. Therefore, the introduction of antimicrobial surface materials may be considered as a potential preventative solution in infection hot spots. Similarly to disinfectants and other hygiene products, the global market for antimicrobial surface coatings is increasing with an annual rate of 10% and is projected to reach USD 7 billion by 2027 [1]. Although other experimental formulations have been used in antimicrobial surfaces, silver, copper, titanium dioxide, and zinc are still the most frequently used active agents [2]. Compared with traditional antibiotics, such metal-based antimicrobial agents have a broad mode of action, which should theoretically prevent the emergence of antimicrobial resistance—a process that has been detected very frequently in the case of antibiotics. Yet, various types of metal-resistance mechanisms in microbes have been described in association to polluted industrial areas and metal mining sites [3]. Furthermore, recent evidence suggests that the appearance of metal resistance may also be linked to the emergence of antibiotic resistance [4], and that such resistant phenotypes may be selected in the presence of sublethal levels of stressors, including various antimicrobial agents [5]. Therefore, ensuring the safety of antimicrobial formulations, and their specific applications in terms of reducing their potential to induce antimicrobial resistance or tolerance, is of great importance when developing sustainable antimicrobial materials. In this work, we propose a strategy to determine the potential of antimicrobial surfaces to induce resistance or tolerance either by enhanced mutation frequency and subsequent selection of resistant mutants or by exchange of genetic material. Along with the fact that such information is required to commercialize biocidal products in the European Union [6], we believe that the proposed framework can be used to ensure the long-term safety and sustainability of antimicrobial surfaces.

Citation: Ivask, A.; Rosenberg, M.; Kisand, V. Towards the Development of Sustainable Antimicrobial Surface Coatings. *Proceedings* **2023**, *92*, 43. https://doi.org/10.3390/proceedings2023092043

Academic Editors: Monika Mortimer, Anne Kahru, Ivo Leito, Riin Rebane and Villem Aruoja

Published: 24 November 2023

Author Contributions: Conceptualization, A.I., V.K. and M.R.; methodology and results, A.I., M.R. and V.K.; project administration, A.I.; funding acquisition, A.I. and V.K. All authors have read and agreed to the published version of the manuscript.

Funding: This research was funded by Estonian Research Council, grant number PRG1496.

Institutional Review Board Statement: Not applicable.

Informed Consent Statement: Not applicable.

Data Availability Statement: Data are available from corresponding author on request.

Conflicts of Interest: The authors declare no conflict of interest.

References

1. Markets and Markets. Antimicrobial Coatings Market. Available online: https://www.marketsandmarkets.com/Market-Reports/antimicrobial-coatings-market-1297.html (accessed on 17 August 2023).
2. Yılmaz, G.E.; Göktürk, I.; Ovezova, M.; Yılmaz, F.; Kılıç, S.; Denizli, A. Antimicrobial Nanomaterials: A Review. *Hygiene* **2023**, *3*, 269–290. [CrossRef]
3. Haferburg, G.; Kothe, E. Microbes and metals: Interactions in the environment. *J. Basic Microbiol.* **2007**, *47*, 453–467. [CrossRef] [PubMed]
4. Vats, P.; Kaur, U.J.; Rishi, P. Heavy metal-induced selection and proliferation of antibiotic resistance: A review. *J. Appl. Microbiol.* **2022**, *132*, 4058–4076. [CrossRef] [PubMed]
5. Andersson, D.I.; Hughes, D. Evolution of antibiotic resistance at non-lethal drug concentrations. *Drug Resist. Updat.* **2012**, *15*, 162–172. [CrossRef] [PubMed]
6. Biocidal Products Directive (Directive 98/8/EC). Available online: https://eur-lex.europa.eu/legal-content/en/ALL/?uri=CELEX%3A31998L0008 (accessed on 17 August 2023).

proceedings

Abstract

Synthesis of Molybdenum Oxide and Sulfide Nanoparticles at Room Temperature Using Organometallic Approach †

Martin Jakoobi [1,*], Guillaume Carnide [1], Laure Vendier [1], Christian Bijani [1], Anne-Françoise Mingotaud [2], Richard Clergereaux [3] and Myrtil L. Kahn [1,*]

[1] Laboratory of Coordination Chemistry (LCC), CNRS UPR 8241, University of Toulouse, 205 Route de Narbonne, 31077 Toulouse, France; guillaume.carnide@lcc-toulouse.fr (G.C.); laure.vendier@lcc-toulouse.fr (L.V.); christian.bijani@lcc-toulouse.fr (C.B.)
[2] Laboratoire des IMRCP, University of Toulouse, CNRS UMR 5623, Paul Sabatier University, 118 Route de Narbonne, CEDEX 09, 31062 Toulouse, France; anne-francoise.mingotaud@cnrs.fr
[3] LAPLACE, CNRS UMR 5213, University of Toulouse, 118 Route de Narbonne, 31062 Toulouse, France; clergereaux@laplace.univ-tlse.fr
* Correspondence: martin.jakoobi@lcc-toulouse.fr (M.J.); myril.kahn@lcc-toulouse.fr (M.L.K.)
† Presented at the International Conference EcoBalt 2023 "Chemicals & Environment", Tallinn, Estonia, 9–11 October 2023.

Keywords: organometallic approach; nanoparticles; mild conditions; size control; colloidal stability; molybdenum oxide; molybdenum sulfide

Citation: Jakoobi, M.; Carnide, G.; Vendier, L.; Bijani, C.; Mingotaud, A.-F.; Clergereaux, R.; Kahn, M.L. Synthesis of Molybdenum Oxide and Sulfide Nanoparticles at Room Temperature Using Organometallic Approach. *Proceedings* **2023**, *92*, 47. https://doi.org/10.3390/proceedings2023092047

Academic Editors: Monika Mortimer, Anne Kahru, Ivo Leito, Riin Rebane and Villem Aruoja

Published: 24 November 2023

In the early 1990s, the seminal work by B. Chaudret and J. S. Bradley led to the establishment of metal nanoparticle (NP) synthesis from organometallic precursors. Since then, this methodology has attracted the attention of many research groups, ours included, and offered possibilities to selectively control the shape and size of these nano-objects under mild reaction conditions. The clever design of substrate and stabilizing ligand systems has allowed the synthesis of late transition metal and their oxide NPs (group 8–12) from precious metals (Ru, Rh, Pd, Pt, Ag, Au) and from first-row (Fe, Co, Ni, Cu, Zn) metal complexes [1,2]. However, to this day, the development of suitable early transition metal complexes (group 3–7) for the preparation of corresponding oxide, sulfide, etc., NPs has attracted only sporadic attention.

This presentation will focus on the use of the highly reactive Mo(0) complex for the synthesis of molybdenum oxide and sulfide NPs at room temperature. These results will illustrate how the choice of the substrate, stabilizing ligand and solvent system can lead to stable colloidal NPs with a diameter of <10 nm in size (Scheme 1).

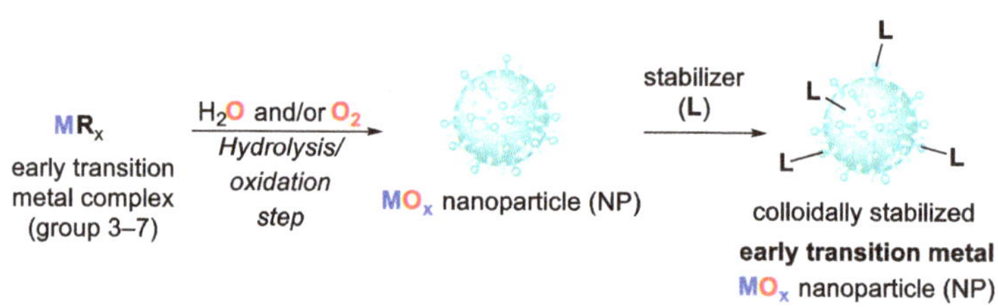

Scheme 1. Synthesis of early transition metal oxide NPs using organometallic approach.

Author Contributions: Conceptualization, M.L.K. and R.C.; writing—original draft preparation, M.J.; formal analysis, M.J., G.C., L.V., C.B. and A.-F.M.; writing—review and editing, M.J. and M.L.K.; supervision M.L.K. All authors have read and agreed to the published version of the manuscript.

Funding: This work was carried out in the framework of the RIF project funded by BPI France.

Institutional Review Board Statement: Not applicable.

Informed Consent Statement: Not applicable.

Data Availability Statement: Not applicable.

Acknowledgments: We thank V. Collière for the TEM measurements.

Conflicts of Interest: The authors declare no conflict of interest.

References

1. Amiens, C.; Chaudret, B.; Ciuculescu-Pradines, D.; Collière, V.; Fajerwerg, K.; Fau, P.; Kahn, M.; Maisonnat, A.; Soulantica, K.; Philippot, K. Organometallic approach for the synthesis of nanostructures. *New J. Chem.* **2013**, *37*, 3374–3401. [CrossRef]
2. Cerezo-Navarrete, C.; Lara, P.; Martínez-Prieto, L.M. Organometallic Nanoparticles Ligated by NHCs: Synthesis, Surface Chemistry and Ligand Effects. *Catalysts* **2020**, *10*, 1144. [CrossRef]

proceedings

Abstract

Neonicotinoids: Agrochemicals with Toxic Impact on Reproductive Functions in Males †

Tomas Jambor *, Lucia Zuscikova, Hana Greifova, Nikola Knizatova, Anton Kovacik and Norbert Lukac

Faculty of Biotechnology and Food Sciences, Institute of Applied Biology, Slovak University of Agriculture in Nitra, Tr. A. Hlinku 2, 949 76 Nitra, Slovakia; xzuscikoval@uniag.sk (L.Z.); hana.greifova@uniag.sk (H.G.); nikola.knizatova@gmail.com (N.K.); anton.kovacik@uniag.sk (A.K.); norbert.lukac@uniag.sk (N.L.)

* Correspondence: tomas.jambor@uniag.sk; Tel.: +421-915160635

† Presented at the International Conference EcoBalt 2023 "Chemicals & Environment", Tallinn, Estonia, 9–11 October 2023.

Keywords: acetamiprid; thiacloprid; leydig cells; Sertoli cells; toxicity

In general, agrochemicals are compounds used to control weeds and diseases in crops during many agronomic practices, and they have become an essential tool in crop protection [1]. Neonicotinoid pesticides are highly effective against some destructive crop pests, and their occurrence in aquatic ecosystems could represent a relevant risk. Acetamiprid N-[(6-chloropyridin-3-yl)methyl]-N′-cyano-N-methylethanimidamide and thiacloprid (2Z)-3-[(6-chloro-3-pyridinyl)methyl]-1,3-thiazolidin-2-ylidenecyanamide) are especially frequently used agrochemicals, with a wide spectrum of efficacy [2]. Currently, exact knowledge about the impact of neonicotinoid exposure on the reproductive system is limited as well as inconsistent. The scientific environment does not provide a relevant background for solving this problem. The objective of our in vitro study was to examine the potential effect of selected neonicotinoids on mouse Sertoli cells. TM4 cells were treated with experimental doses of acetamiprid (10 to 500 μM) and thiacloprid (7.8 to 500 μM) for 48 hours of exposure. Metabolic activity and cell membrane integrity were examined to determine the potential toxicity. The results of an alamarblue assay revealed that higher experimental doses of acetamiprid (200–500 μM) significantly ($p < 0.0001$) decreased the metabolic activity of exposed TM4 Sertoli cells. A similar tendency was confirmed after thiacloprid exposure when significant ($p < 0.0001$) cytotoxicity started from 125 to 500 μM. The cell membrane integrity, evaluated via a CFDA-AM assay, showed a significant ($p < 0.01$) decrease at 250 or 300 μM followed by significant ($p < 0.001$; $p < 0.0001$) inhibition at 350 and 500 μM of acetamiprid. In the case of thiacloprid, the presented parameter was significantly ($p < 0.01$) inhibited at 125 and 250 μM, while the highest concentrations, 300 and 500 μM, caused significant changes ($p < 0.001$; $p < 0.0001$). Considerably more detailed and systematic research on thiacloprid toxicology is definitely required for a better understanding of the risks associated with reproductive health.

Author Contributions: Conceptualization, T.J. and N.L.; methodology, T.J. and H.G.; validation, N.L. and A.K.; formal analysis, N.K.; original draft preparation, T.J. and L.Z.; writing—review and editing, N.L.; supervision, N.L. and A.K.; project administration, T.J., A.K. and N.L; funding acquisition, N.L. All authors have read and agreed to the published version of the manuscript.

Funding: This research was funded by the Scientific Agency of the Slovak Republic (VEGA), No. 1/0083/21, and the Slovak Research and Development Agency, Grant No. APVV-21-0168 and APVV-20-0218.

Institutional Review Board Statement: Not applicable.

Informed Consent Statement: Not applicable.

Citation: Jambor, T.; Zuscikova, L.; Greifova, H.; Knizatova, N.; Kovacik, A.; Lukac, N. Neonicotinoids: Agrochemicals with Toxic Impact on Reproductive Functions in Males. *Proceedings* **2023**, *92*, 20. https://doi.org/10.3390/proceedings2023092020

Academic Editors: Monika Mortimer, Anne Kahru, Ivo Leito, Riin Rebane and Villem Aruoja

Published: 21 November 2023

Data Availability Statement: Data sharing not applicable.

Conflicts of Interest: The authors declare no conflict of interest.

References

1. Gupta, S.; Khanna, R. Agrochemicals as a Potential Cause of Ground Water Pollution: A Review. *Int. J. Chem. Stud.* **2018**, *6*, 985–990.
2. Renaud, M.; Akeju, T.; Natal-da-Luz, T.; Leston, S.; Rosa, J.; Ramos, F.; Sousa, J.P.; Azevedo-Pereira, H.M.V.S. Effects of the Neonicotinoids Acetamiprid and Thiacloprid in Their Commercial Formulations on Soil Fauna. *Chemosphere* **2018**, *194*, 85–93. [CrossRef] [PubMed]

Abstract

Liquid Complex Fertilizers with Bio-Additives †

Kristina Jančaitienė * and Ramunė Sidaraitė

Department of Physical and Inorganic Chemistry, Kaunas University of Technology, Radvilėnų rd. 19, 50254 Kaunas, Lithuania; ramune.sidaraite@gmail.com
* Correspondence: kristina.jancaitiene@ktu.lt

† Presented at the International Conference EcoBalt 2023 "Chemicals & Environment", Tallinn, Estonia, 9–11 October 2023.

Keywords: liquid fertilizers; nitrogen; bio fertilizers; blue-pod lupin

Citation: Jančaitienė, K.; Sidaraitė, R. Liquid Complex Fertilizers with Bio-Additives. *Proceedings* **2023**, *92*, 30. https://doi.org/10.3390/proceedings2023092030

Academic Editors: Monika Mortimer, Anne Kahru, Ivo Leito, Riin Rebane and Villem Aruoja

Published: 22 November 2023

Every year, approximately 2.5 billion tons of waste are generated in the European Union. Due to these reasons and the limited amount of raw materials, the integration of the circular economy is encouraged [1]. The fertilizer industry is one of the industries that is characterized by the recycling of by-products into new products. One of the by-products in the fertilizer industry is the liquid phase, which is obtained as a secondary product after the crystallization and filtration of the solid primary product. Liquid fertilizers have many advantages: they are highly compatible with trace elements, fungicides, physiologically active substances or other important additives [2]. It is very important to choose the right fertilizer, because plants need to be given the required amount of nutrients. If at least one element is missing, the plant may rot or even die [3]. During this work, the possibility of using the post-crystallization solution for the production of liquid complex fertilizers was examined. To achieve this goal, the concentrations of plant nutrients and chlorine in the post-crystallization solution and its chemical—physical properties were determined. Considering the need to increase the concentration of nitrogen in the post-crystallization solution, studies were conducted, during which the influence of different nitrogen compounds on the crystallization temperature of the post-crystallization solution were observed. Studies have also been carried out in which solutions (as organic nitrogen additives) that would be highly compatible with the post-crystallization solution were extracted from the lupine (*Lupinus polyphyllus*). After the chemical analysis, the liquid complex fertilizers were found to contain 1.4%–P_2O_5, 4%–Cl and approximately 13.5%–K_2O. Depending on the nitrogen additions, the concentration of nitrogen in the liquid fertilizers varied from 1.02% to 3.78% N. Four liquid complex fertilizers of different brands were obtained and used: 1-2-14; 3-2-14+ME; 4-2-14+ME+mineralized lupine leaves; and 4-2-14+ME+lupine leaf extract.

Author Contributions: Conceptualization, K.J.; methodology, K.J. and R.S.; software, K.J.; resources, R.S.; data curation, K.J. and R.S.; writing—original draft preparation, K.J.; writing—review and editing, K.J.; visualization, K.J.; supervision, K.J. All authors have read and agreed to the published version of the manuscript.

Funding: This research received no external funding.

Institutional Review Board Statement: Not applicable.

Informed Consent Statement: Not applicable.

Data Availability Statement: Data are contained within the article.

Conflicts of Interest: The authors declare no conflict of interest.

References

1. European Parliament (EP). Circular Economy: Definition, Importance and Benefits. Available online: https://www.eumonitor.eu/9353000/1/j9vvik7m1c3gyxp/vknegugz7hwu?ctx=vjxzjv7ta8z1 (accessed on 27 February 2023).
2. Bommaraju, T.V.; Barnicki, S.D.; Kent, J.A. *Handbook of Industrial Chemistry and Biotechnology*, 3rd ed.; Springer: New York, NY, USA, 2017; pp. 784–785.
3. Paleckienė, R.; Sviklas, M.A. *Trąšų Agrochemija: Metodinė Priemonė, Vilnius*; KTU Faculty of Chemical Technology: Kaunas, Lithuania, 2012; pp. 32–107.

proceedings

Abstract

Selective Capture of Small Environmental Pollutants by Cyclohexanohemicucurbit[*n*]urils †

Tatsiana Jarg, Ketren-Marlein Lootus, Sandra Kaabel and Riina Aav *

Department of Chemistry and Biotechnology, Tallinn University of Technology, Akadeemia tee 15, 12618 Tallinn, Estonia
* Correspondence: riina.aav@taltech.ee

† Presented at the International Conference EcoBalt 2023 "Chemicals & Environment", Tallinn, Estonia, 9–11 October 2023.

Keywords: hemicucurbituril; heterocycle; haloalkane; pollutant; inclusion complex; solid-phase extraction; sorbent recycling

Persistent organic pollutants that occur in different environmental matrices create a need for their sensing, removal and remediation. Cyclohexanochemicucurbit[*n*]urils (cycHC[*n*], *n* = 6 or 8) are chiral macrocyclic cavitands with a well-defined electron-deficient cavity suitable for the accommodation of various electron-rich guests [1–3]. The current work describes the application of cycHC[8] as a molecular container for the encapsulation of small neutral organic molecules, including environmental pollutants (mustard gas degradation products, haloalkanes) and biologically active compounds. The formation of inclusion complexes in solid state was confirmed via SC-XRD. Complexation studies in solution through 1H NMR and ITC revealed host–guest interactions between cycHC[8] and hydrophobic S- and O-containing neutral heterocycles, as well as small haloalkanes, in methanol and methanol–water media, resulting in stronger binding upon an increase in solvent polarity. Since the macrocycle is insoluble in water, it was further utilized for the selective capture of the neutral guests from aqueous solutions [3].

Citation: Jarg, T.; Lootus, K.-M.; Kaabel, S.; Aav, R. Selective Capture of Small Environmental Pollutants by Cyclohexanohemicucurbit[*n*]urils. *Proceedings* **2023**, *92*, 85. https://doi.org/10.3390/proceedings2023092085

Academic Editors: Monika Mortimer, Anne Kahru, Ivo Leito, Riin Rebane and Villem Aruoja

Published: 29 January 2024

Author Contributions: Conceptualization, supervision and methodology, R.A.; validation, T.J.; formal analysis and investigation, T.J., K.-M.L. and S.K. All authors have read and agreed to the published version of the manuscript.

Funding: This research was funded by Estonian Research Council Grants (PRG399, MOBJD556 and MOBJD592), European Regional Development Fund (CoE 2014-2020.4.01.15-0013 and CoE TK134), and H2020-FETOPEN 828779 (INITIO).

Institutional Review Board Statement: Not applicable.

Informed Consent Statement: Not applicable.

Data Availability Statement: Not applicable.

Conflicts of Interest: The authors declare no conflicts of interest.

References

1. Öeren, M.; Shmatova, E.; Tamm, T.; Aav, R. Computational and Ion Mobility MS Study of (*All-S*)-Cyclohexylhemicucurbit[6]uril Structure and Complexes. *Phys. Chem. Chem. Phys.* **2014**, *16*, 19198–19205. [CrossRef] [PubMed]
2. Kaabel, S.; Adamson, J.; Topić, F.; Kiesilä, A.; Kalenius, E.; Öeren, M.; Reimund, M.; Prigorchenko, E.; Lõokene, A.; Reich, H.J.; et al. Chiral Hemicucurbit[8]uril as an Anion Receptor: Selectivity to Size, Shape and Charge Distribution. *Chem. Sci. J.* **2017**, *8*, 2184–2190. [CrossRef] [PubMed]
3. Shalima, T.; Mishra, K.A.; Kaabel, S.; Ustrnul, L.; Bartkova, S.; Tõnsuaadu, K.; Heinmaa, I.; Aav, R. Cyclohexanohemicucurbit[8]uril Inclusion Complexes With Heterocycles and Selective Extraction of Sulfur Compounds From Water. *Front. Chem.* **2021**, *9*, 1036. [CrossRef] [PubMed]

proceedings

Abstract

Interactions of Diatom *Cyclotella meneghiniana* and Citrate Coated Silver Nanoparticles †

Arin Kantarciyan *, Inés Segovia-Campos and Vera I. Slaveykova

Department F.-A. Forel for Environmental and Aquatic Sciences, University of Geneva, 66 Boulevard Carl-Vogt, CH-1211 Geneva, Switzerland; ines.segoviacampos@unige.ch (I.S.-C.); vera.slaveykova@unige.ch (V.I.S.)
* Correspondence: arin.kantarciyan@unige.ch; Tel.: +41-22-379-03-86
† Presented at the International Conference EcoBalt 2023 "Chemicals & Environment", Tallinn, Estonia, 9–11 October 2023.

Keywords: nanoparticles; toxicity; diatom

Citation: Kantarciyan, A.; Segovia-Campos, I.; Slaveykova, V.I. Interactions of Diatom *Cyclotella meneghiniana* and Citrate Coated Silver Nanoparticles. *Proceedings* **2023**, *92*, 49. https://doi.org/10.3390/proceedings2023092049

Academic Editor: Monika Mortimer, Anne Kahru, Ivo Leito, Riin Rebane and Villem Aruoja

Published: 24 November 2023

Silver nanoparticles (AgNPs) are extensively utilized engineered nanomaterials that inevitably find their way into the aquatic environment [1]. A significant amount of research has been conducted to assess their potential toxicity to aquatic biota [2]. However, the underlying cellular mechanisms involved in the toxicity and tolerance of diatoms to AgNPs are still poorly understood. The present work aimed to gain better insight into the response of diatom *Cyclotella meneghiniana* to AgNP exposure, as a representative model for the lower trophic organisms in a freshwater environment, and its underlying mechanisms. *C. meneghiniana* was exposed to various concentrations of Cit-AgNPs (ranging from 0.001 mg/L to 5 mg/L) for up to 72 h, and the biological responses were compared with those induced by dissolved Ag^+. The response of diatoms to Cit-AgNP and Ag^+ was characterized in terms of diatom growth, membrane permeability, photosynthesis alterations, and morphological changes. The stability of the Cit-AgNPs in the exposure medium was also investigated by determining their dissolution, surface charge and hydrodynamic size. The DLS and SPR-UV-vis results showed a shift in the size distribution of cit-AgNPs towards higher values, which was related to aggregation/agglomeration processes. The dissolution of cit-AgNP in the exposure media increased over time and was concentration-dependent. The calculated 72 h-EC50 values, based on growth inhibition, were 0.348 ± 0.038 mg/L and 0.019 ± 0.001 mg/L for cit-AgNP and Ag^+, respectively, suggesting a higher toxicity of Ag^+ compared to cit-AgNP for *C.meneghiniana*. Short-term exposure (24 h) to Cit-AgNPs and Ag^+ resulted in reduced chlorophyll autofluorescence and impaired membrane integrity in *C. meneghiniana*. Furthermore, the photosystem II was affected, as indicated by a decrease in the quantum yield (Fv/Fm) and an increase in non-photochemical quenching (NPQ). Cells exposed to Cit-AgNPs and Ag^+ exhibited higher levels of proline accumulation compared to the control, implying an activation of the antioxidant mechanisms in diatom, since proline plays a role in ROS scavenging [3]. Additionally, the SEM-EDS analysis revealed an increased presence of polyphosphate bodies (PPB) in both the Cit-AgNP- and Ag^+-treated cells, in response to metal toxicity and stress. Indeed, polyphosphates are known as a chelator of cations, and their accumulation is linked to abiotic stress [4]. This study demonstrates that synergistic mechanisms are adopted by *C. meneghiniana* to deal with toxic levels of silver in both its ionic and nanoparticulate forms.

Author Contributions: Conceptualization, A.K. and V.I.S.; methodology, A.K. and V.I.S.; software, A.K. and V.I.S.; validation, A.K. and V.I.S.; formal analysis, A.K and I.S.-C.; investigation, A.K.; resources, A.K. and V.I.S.; writing—original draft preparation, A.K., I.S.-C. and V.I.S.; writing—review and editing, A.K., I.S.-C. and V.I.S.; visualization, A.K. and V.I.S.; supervision, V.I.S.; project administration, V.I.S.; funding acquisition, V.I.S. All authors have read and agreed to the published version of the manuscript.

Funding: This research was funded by Swiss National Science Foundation Project No. 204174.

Institutional Review Board Statement: Not applicable.

Informed Consent Statement: Not applicable.

Data Availability Statement: The data presented in this study are available upon request from the corresponding author.

Acknowledgments: The author wants to thank Rocco Gasco for fruitful scientific discussions.

Conflicts of Interest: The authors declare no conflict of interest.

References

1. McGillicuddy, E.; Murray, I.; Kavanagh, S.; Morrison, L.; Fogarty, A.; Cormican, M. Silver nanoparticles in the environment: Sources, detection and ecotoxicology. *Sci. Total Environ.* **2017**, *575*, 231–246. [CrossRef] [PubMed]
2. Kalantzi, I.; Mylona, K.; Toncelli, C.; Bucheli, T.D.; Knauer, K.; Pergantis, S.A. Ecotoxicity of silver nanoparticles on plankton organisms: A review. *J. Nanoparticle Res.* **2019**, *21*, 65. [CrossRef]
3. Torres, M.A.; Barros, M.P.; Campos, S.C.G.; Pinto, E.; Rajamani, S.; Sayre, R.T. Biochemical biomarkers in algae and marine pollution: A review. *Ecotoxicol. Environ. Saf.* **2008**, *71*, 1–15. [CrossRef] [PubMed]
4. Sanz-Luque, E.; Bhaya, D.; Grossman, A.R. Polyphosphate: A Multifunctional Metabolite in Cyanobacteria and Algae. *Front. Plant Sci.* **2020**, *11*, 938. [CrossRef] [PubMed]

proceedings

MDPI

Abstract

Synthesis and Antibacterial Efficiency of Chitosan–Copper Oxide Nanocomposites †

Jüri Laanoja [1,2], Mariliis Sihtmäe [1], Maarja Otsus [1], Anne Kahru [1] and Kaja Kasemets [1,*]

1 Laboratory of Environmental Toxicology, National Institute of Chemical Physics and Biophysics, 12618 Tallinn, Estonia; juri.laanoja@gmail.com (J.L.); sihtmae@gmail.com (M.S.); maarja.otsus@gmail.com (M.O.); anne.kahru@kbfi.ee (A.K.)
2 School of Science, Tallinn University of Technology, 19086 Tallinn, Estonia
* Correspondence: kaja.kasemets@kbfi.ee
† Presented at the International Conference EcoBalt 2023 "Chemicals & Environment", Tallinn, Estonia, 9–11 October 2023.

Keywords: antimicrobial resistance; bacteria; chitosan; copper; copper oxide; nanocomposites; nanoparticles; nanotechnology

Citation: Laanoja, J.; Sihtmäe, M.; Otsus, M.; Kahru, A.; Kasemets, K. Synthesis and Antibacterial Efficiency of Chitosan–Copper Oxide Nanocomposites. *Proceedings* **2023**, *92*, 35. https://doi.org/10.3390/proceedings2023092035

Academic Editors: Monika Mortimer, Ivo Leito, Riin Rebane and Villem Aruoja

Published: 22 November 2023

Antimicrobial resistance is among the most serious global healthcare problems today. In Europe, a third of the estimated 8.9 million yearly hospital-acquired infections in 2016 and 2017 were caused by antibiotic-resistant bacteria [1]. One possible source for novel, efficient antimicrobials is nanotechnology. Some metallic nanoparticles (NPs) like silver, CuO, and ZnO are already commercially available for biomedical applications. NPs are often coated with different polymers to improve their characteristics. An intriguing biopolymer to combine with metal NPs is the antimicrobial, biocompatible, and immunomodulating chitosan, presently used in bandages and wound dressings. In this study, novel chitosan–CuO nanocomposites (NCs) were synthesized via the precipitation of copper acetate by sodium hydroxide in the presence of chitosan (50–190 kDa, Sigma Aldrich, Schnelldorf, Germany). Varying copper to chitosan weight ratios (1:0.3, 1:1, and 1:3) were used. The NCs were characterized by DLS, EDX, FTIR, SEM, XPS, and XRD. The minimum bactericidal concentrations (MBC) against clinically relevant gram-negative (*Escherichia coli* ATCC 25922, *E. coli* MG1655, *Pseudomonas aeruginosa* ATCC 27853) and gram-positive (*Staphylococcus aureus* ATCC 6538) bacteria were determined by the Spot test [2]. Flow cytometry and confocal laser scanning microscopy were used to reflect the interactions between NCs and bacteria. The zeta potential of the synthesized NCs was >40 mV. The NCs' hydrodynamic diameter and polydispersity index increased with higher chitosan content, varying within 90–180 nm and 0.24–0.33, respectively. Based on the XRD analysis, the CuO portion of the NCs had a crystalline structure. The NCs were similarly effective against gram-negative and -positive bacteria, displaying MBC values of 0.13–0.25 mg Cu/L after 24 h of exposure. Interestingly, after a 1-h period of exposure, the NCs were more toxic against gram-negative bacteria than Cu ions, suggesting that chitosan may facilitate the interaction of NCs and bacterial cells, enabling the immediate shedding of Cu ions alongside the bacteria. Furthermore, after a 1-h exposure, the NCs with a higher chitosan content were up to two times more biocidal against gram negative bacteria than NCs with a lower chitosan concentration.

Author Contributions: Conceptualization, J.L. and K.K.; methodology, formal analysis and investigation, J.L., M.S., M.O. and K.K.; writing, J.L.; review and editing, M.S., A.K. and K.K.; supervision, project administration, and funding acquisition, K.K. All authors have read and agreed to the published version of the manuscript.

Funding: This work was supported by the Estonian Research Council project PRG749 and conducted using the NAMUR+ core facility (TT13).

Institutional Review Board Statement: Not applicable.

Informed Consent Statement: Not applicable.

Data Availability Statement: Not applicable.

Acknowledgments: The authors thank Svetlana Vihodceva, Raivis Eglītis, Mairis Iesalnieks, and Inna Juhņeviča from Riga Technical University for their invaluable help.

Conflicts of Interest: The authors declare no conflict of interest.

References

1. European Centre for Disease Prevention and Control. Available online: https://www.ecdc.europa.eu/en/publications-data/infographic-healthcare-associated-infections-threat-patient-safety-europe (accessed on 22 June 2023).
2. Suppi, S.; Kasemets, K.; Ivask, A.; Künnis-Beres, K.; Sihtmäe, M.; Kurvet, I.; Aruoja, V.; Kahru, A. A novel method for comparison of biocidal properties of nanomaterials to bacteria, yeasts and algae. *J. Hazard Mater.* **2015**, *286*, 75–84. [CrossRef] [PubMed]

proceedings

Abstract

Multi-Element Profile Characterization in Monofloral Honey †

Kriss Davids Labsvards [1,2,*], Anastasija Borisova [2], Kristina Kokina [2], Maris Bertins [1], Arturs Viksna [1] and Vita Rudovica [1]

1 Department of Chemistry, University of Latvia, Jelgavas Street 1, LV-1004 Riga, Latvia; maris.bertins@lu.lv (M.B.); arturs.viksna@lu.lv (A.V.); vita.rudovica@lu.lv (V.R.)
2 Institute of Food Safety, Animal Health and Environment "BIOR", Lejupes Street 3, LV-1076 Riga, Latvia; anastasija.borisova@bior.lv (A.B.); kristina.kokina@bior.lv (K.K.)
* Correspondence: kriss_davids.labsvards@lu.lv; Tel.: +371-26395784

† Presented at the International Conference EcoBalt 2023 "Chemicals & Environment", Tallinn, Estonia, 9–11 October 2023.

Keywords: honey; ICP-MS; floral origins; macro elements; trace elements; chemical profile

Apiculture, in general, is an important sector of the national economy due to the environmental benefits of pollination. Honey is the main product of apiculture and is regarded as a product with health-promoting properties; it is therefore often used for medical purposes [1]. The high demand for honey raises concerns about the honey quality available in the market because of the possible presence of counterfeited or fraudulent products, thereby highlighting the need for modern instrumental analysis. Monofloral honey is gathered from the majority of single floral sources, thus leading to a unique taste, and organoleptic or visual properties, which consumers might find more attractive than regular polyfloral honey. Determining the macro and trace element profile of these honeys is valuable, serving a purpose to consumers, manufacturers, and researchers. In this study, 83 honey samples of different floral origins were used, and the floral origins were confirmed via melissopalynology analysis. The macro and trace element profile of the honey was determined using inductively coupled plasma–mass spectrometry (ICP-MS). The results were processed using chemometric methods, including principal component analysis (PCA) and a dendrogram of hierarchical clustering. Significant differences were determined using one-way ANOVA analysis and Fisher's test. In total, 30 different elements were found in natural honey of Latvian origins, and the potential use of 18 of them as floral markers for buckwheat, clover, heather, linden, rapeseed and willow honey was determined. Heather honey showed the most diverse element profile, with increased concentrations of As, Ba, Ca, Cs, Fe, K, Mn, Rb and Tl. Compared to other studies, the preliminary results of the ICP-MS show one of the most versatile evaluations of floral origin [2]. The preliminary results regarding macro and trace elements show their potential use as biomarkers for the evaluation of honey floral origin and the evaluation of concentrations of elements harmful to health.

Citation: Labsvards, K.D.; Borisova, A.; Kokina, K.; Bertins, M.; Viksna, A.; Rudovica, V. Multi-Element Profile Characterization in Monofloral Honey. *Proceedings* **2023**, *92*, 31. https://doi.org/10.3390/proceedings2023092031

Academic Editors: Monika Mortimer, Anne Kahru, Ivo Leito, Riin Rebane and Villem Aruoja

Published: 22 November 2023

Author Contributions: Conceptualization, K.D.L. and A.V.; methodology, K.D.L., K.K. and V.R.; software, K.D.L.; validation, A.B., K.D.L. and M.B.; formal analysis, A.B. and M.B.; investigation, K.D.L., A.B. and A.V.; resources, K.D.L. and A.V.; data curation, K.D.L., M.B., A.B. and V.R.; writing—original draft preparation, K.D.L.; writing—review and editing, K.D.L., A.V., K.K. and V.R.; visualization, K.D.L. and V.R.; supervision, K.D.L., A.V., K.K. and V.R.; project administration, K.D.L.; funding acquisition, K.D.L. All authors have read and agreed to the published version of the manuscript.

Funding: "Strengthening of the capacity of doctoral studies at the University of Latvia within the framework of the new doctoral model", identification No. 8.2.2.0/20/I/006.

Institutional Review Board Statement: Not applicable.

Informed Consent Statement: Not applicable.

Data Availability Statement: The data presented in this study are available upon request from the corresponding author.

Acknowledgments: The authors thank the Latvian Beekeeping Association for the assessment of honey floral origins using melissopalynology analysis.

Conflicts of Interest: The authors declare no conflict of interest. The funders had no role in the design of the study; in the collection, analyses, or interpretation of data; in the writing of the manuscript; or in the decision to publish the results.

References

1. Skadiņš, I.; Labsvārds, K.D.; Grava, A.; Amirian, J.; Tomsone, L.E.; Ruško, J.; Viksna, A.; Bandere, D.; Brangule, A. Antimicrobial and Antibiofilm Properties of Latvian Honey against Causative Agents of Wound Infections. *Antibiotics* **2023**, *12*, 816. [CrossRef]
2. Labsvards, K.D.; Rudovica, V.; Kluga, R.; Rusko, J.; Busa, L.; Bertins, M.; Eglite, I.; Naumenko, J.; Salajeva, M.; Viksna, A. Determination of Floral Origin Markers of Latvian Honey by Using IRMS, UHPLC-HRMS, and 1H-NMR. *Foods* **2022**, *11*, 42. [CrossRef]

proceedings

MDPI

Abstract

Future Perspective of In Situ Soil Analysis †

Mari-Liis Leinus [1,*], Piia Jõul [1], Martin Růžička [1], Jelena Gorbatšova [1], Merike Vaher [1], Ruth Shimmo [2] and Jekaterina Mazina-Šinkar [1]

1 Department of Chemistry and Biotechnology, Tallinn University of Technology, 12618 Tallinn, Estonia
2 School of Natural Sciences and Health, Tallinn University, 10120 Tallinn, Estonia
* Correspondence: mari-liis.leinus@taltech.ee
† Presented at the International Conference EcoBalt 2023 "Chemicals & Environment", Tallinn, Estonia, 9–11 October 2023.

Keywords: soil health; soil fertility; nutrients; portable capillary electrophoresis; GHG emissions; fertilization of soils; over fertilization

Soil health monitoring is becoming progressively more important to meet the needs of sustainable agriculture and climate policy goals. Portable capillary electrophoresis (CE) devices could be used to monitor micro and macro nutrients available to plants in soil in real time as anions and cations to make informed decisions about the need for fertilization. These devices could be considered precise, cheap and fast alternatives to time-consuming laboratory analyses and sensors that are currently on the market that can determine only N, P and K, and not other important macro or micro elements [1–5]. Method development for determining cations in soil samples was carried out using commercial CE device coupled to a contactless conductivity detector (C4D) with the intent to later implement the developed method on a portable CE device. The separation of eight cations (NH_4^+, K^+, Na^+, Ca^{2+}, Mg^{2+}, Mn^{2+}, Zn^{2+} and Cu^{2+}) was achieved using a capillary with an effective length of 47.5 cm and an inner diameter of 50 µm. Briefly, a 6 M acetic acid solution was used as a background electrolyte, and we used an applied voltage of 15 kV. The duration of analysis was 20 min (including 3 min of pre-washing). Linearity for analytes was determined to be in the range of 1–10 mM, which encompasses the limit of detection (LoD) (0.1–0.9 mM) and limit of quantification (LoQ) (0.3–1.8 mM) for the analytes. The extraction of cations was tested from 1 min up to 24 h using distilled water and 0.01 M $CaCl_2$ [6,7]. No remarkable differences in extraction recoveries were observed within the tested timeframe. Therefore, 1–3 min was suggested as an optimum extraction time for the in situ extraction procedure. The developed analysis method is suitable for the qualitative and quantitative analysis of eight cations extracted from soil and can be implemented on portable CE devices. Further optimization of simple and quick sample preparation must be carried out so that it could be easily used directly on the field prior to analysis.

Citation: Leinus, M.-L.; Jõul, P.; Růžička, M.; Gorbatšova, J.; Vaher, M.; Shimmo, R.; Mazina-Šinkar, J. Future Perspective of In Situ Soil Analysis. *Proceedings* **2023**, *92*, 10. https://doi.org/10.3390/proceedings2023092010

Academic Editors: Monika Mortimer, Anne Kahru, Ivo Leito, Riin Rebane and Villem Aruoja

Published: 21 November 2023

Author Contributions: Conceptualization, J.G., M.V. and J.M.-Š.; methodology, M.-L.L. and P.J.; validation, M.-L.L., P.J. and M.V.; hardware, M.R., J.G. and J.M.-Š.; formal analysis, M.-L.L. and P.J.; investigation, M.-L.L. and P.J.; resources, J.G. and J.M.-Š.; data curation, M.-L.L. and P.J.; writing—original draft preparation, M.-L.L.; writing—review and editing, R.S. and J.M.-Š.; visualization, M.-L.L. and M.R.; supervision, R.S. and J.M.-Š.; project administration, J.M.-Š.; funding acquisition, J.M.-Š. All authors have read and agreed to the published version of the manuscript.

Funding: This research was funded by Environmental Investment Centre (KIK), grant number Kliima.3.01.22-0101 ("SmartAGRO"), and Estonian Research Council, grant number MOBJD1015.

Institutional Review Board Statement: Not applicable.

Informed Consent Statement: Not applicable.

Data Availability Statement: The data presented in this study are available upon request from the corresponding author.

Acknowledgments: Our partners in Ukraine, National University of Life and Environmental Sciences of Ukraine, Agriculture (Farming) Enterprise "Sophiya S+", and Agriculture (Farming) Enterprise "Roksolana", are acknowledged for providing soil samples.

Conflicts of Interest: The authors declare no conflict of interest. The funders had no role in the design of the study; in the collection, analyses, or interpretation of data; in the writing of the manuscript; or in the decision to publish the results.

References

1. Environmental XPRT, Renke—Model RS-NPK-TR—Soil NPK Sensor. Available online: https://www.environmental-expert.com/products/model-rs-npk-tr-soil-npk-sensor-784865 (accessed on 31 May 2023).
2. Teralytic Soil Probe. Available online: https://teralytic.com/soil-probe/ (accessed on 31 May 2023).
3. Paul-Tech Soil Station Service. Available online: https://paul-tech.com/paul-tech-soil-station-service/ (accessed on 31 May 2023).
4. AgroCares Nutrient Scanner. Available online: https://www.agrocares.com/scanners/ (accessed on 31 May 2023).
5. Bruker, Application Note N526, Online Quality Control of Mineral Fertilizers Using FT-NIR Spectroscopy. Available online: https://www.bruker.com/en/meta/forms/bopt-form-pages/thank-you-an-n526-form/_jcr_content/root/contentpar/twocolumns_copy_copy/contentpar-1/calltoaction.download-asset.pdf/primaryButton/AN_N526_Fertilizer_online_EN.pdf (accessed on 31 May 2023).
6. Pontes, F.V.M.; Carneiro, M.C.; Vaitsman, D.S.; Monteiro, M.I.C.; da Silva, L.I.D.; de Souza, E.D.M.F.; Neto, A.A. Fast and simultaneous ultrasound-assisted extraction of exchangeable-NH_4^+, NO_3^- and NO_2^- species from soils followed by ion chromatography determination. *Chem. Speciat. Bioavailab.* **2012**, *24*, 229. [CrossRef]
7. Stanišić, S.M.; Ignjatović, L.M.; Manojlović, D.; Dojčinović, B. The comparison of sample extraction procedures for the determination of cations in soil by IC and ICP-AES. *Cent. Eur. J. Chem.* **2011**, *9*, 481–484. [CrossRef]

proceedings

MDPI

Abstract

"Hook & Loop" Interactions between Fibrous Microplastics and Zooplankton †

Cuizhu Ma [1,2,*] **and Vera I. Slaveykova** [1]

[1] Department F.-A. Forel for Environmental and Aquatic Sciences, Environmental Biogeochemistry and Ecotoxicology, Faculty of Sciences, Earth and Environment Sciences, University of Geneva, 66 Boulevard Carl-Vogt, CH-1211 Geneva, Switzerland
[2] State Key Laboratory of Estuarine and Coastal Research, East China Normal University, Shanghai 200241, China
* Correspondence: cuizhu.ma@etu.unige.ch; Tel.: +41-22-379-0339

† Presented at the International Conference EcoBalt 2023 "Chemicals & Environment", Tallinn, Estonia, 9–11 October 2023.

Keywords: hook and loop; microfibers; *Daphnia magna*

Citation: Ma, C.; Slaveykova, V.I. "Hook & Loop" Interactions between Fibrous Microplastics and Zooplankton. *Proceedings* **2023**, *92*, 8. https://doi.org/10.3390/proceedings2023092008

Academic Editors: Monika Mortimer, Anne Kahru, Ivo Leito, Riin Rebane and Villem Aruoja

Published: 21 November 2023

Today, microplastics (1 μm–5 mm) have been found within organisms ranging from small invertebrates to large mammals [1–3]. However, fibrous microplastics, the dominant form of microplastics in the environment, are still not the focus of research [4]. Previous studies indicated that microfibers caused more behavioral toxicity in organisms than pellets and fragments because of organism entanglement [5,6]. Here, the overarching research objective was to gain more information about the behavioral effects of fibrous microplastics on freshwater zooplankton *Daphnia magna*. *D. magna* were exposed to fibrous polyester microplastics sized 300 ± 192 μm at different concentrations (10, 10^2, and 10^3 items/mL) in the presence of algal food. Behavioral responses, particularly swimming speed and swimming trajectory, were recorded and analyzed via a tracker for 7 days. The frequency of hop and sink behavior decreased when *D. magna* were exposed to microfibers. Moreover, a special phenomenon, namely "hook & loop", was noticed when examining the connection between fibrous microplastics and *D. magna*, especially in the 10^2 items/mL group. Microfibers can easily twin the antennae and tail claws of *D. magna*, which is due to their faint pectinate spines. The strong effect of microfibers intertwined with algae or impurities prevented *D. magna* from moving freely, which caused a decrease in swimming speed and swimming trajectories. More interestingly, this phenomenon was not dose-dependent. *D. magna* exposed to much higher concentrations (10^3 items/mL) swam freely because of the aggregation of microplastics themselves, which indicates the importance of microplastic concentration number. Overall, this study demonstrates that "hook & loop" interactions between fibrous microplastics and zooplankton could be directly causing the organisms' behavior changes, which indicates that the interactions of microplastics and organisms play a key role in understanding the behavior, fate, and effects of microplastics in the aquatic environment.

Author Contributions: Conceptualization, C.M. and V.I.S.; methodology, C.M. and V.I.S.; software, C.M. and V.I.S.; validation, C.M. and V.I.S.; formal analysis, C.M.; investigation, C.M.; resources, C.M. and V.I.S.; data curation, C.M. and V.I.S.; writing—original draft preparation, C.M. and V.I.S.; writing—review and editing, C.M. and V.I.S.; visualization, C.M. and V.I.S.; supervision, V.I.S.; project administration, V.I.S.; funding acquisition, V.I.S. All authors have read and agreed to the published version of the manuscript.

Funding: This research was funded by the China Scholarship Council (20220716) and the University of Geneva.

Institutional Review Board Statement: Not applicable.

Informed Consent Statement: Not applicable.

Data Availability Statement: The data presented in this study are available upon request from the corresponding author.

Acknowledgments: The authors thank Huahong Shi for his informative suggestions and for providing fibrous microplastics. The authors also thank Marlita Marlita for her help with the culture of zooplankton and algae food.

Conflicts of Interest: The authors declare no conflict of interest.

References

1. Cole, M.; Lindeque, P.; Fileman, E.; Halsband, C.; Goodhead, R.; Moger, J.; Galloway, T.S. Microplastic ingestion by zooplankton. *Environ. Sci. Technol.* **2013**, *47*, 6646–6655. [CrossRef] [PubMed]
2. Rebolledo, E.L.B.; Van Franeker, J.A.; Jansen, O.E.; Brasseur, S.M.J.M. Plastic ingestion by harbour seals (*Phoca vitulina*) in the Netherlands. *Mar. Pollut. Bull.* **2013**, *67*, 200–202. [CrossRef] [PubMed]
3. Sanchez, W.; Bender, C.; Porcher, J.M. Wild gudgeons (*Gobio gobio*) from French rivers are contaminated by microplastics: Preliminary study and first evidence. *Environ. Res.* **2014**, *128*, 98–100. [CrossRef] [PubMed]
4. Rebelein, A.; Int-Veen, I.; Kammann, U.; Scharsack, J.P. Microplastic fibers—Underestimated threat to aquatic organisms? *Sci. Total Environ.* **2021**, *777*, 146045. [CrossRef] [PubMed]
5. Gray, A.D.; Weinstein, J.E. Size- and shape-dependent effects of microplastic particles on adult dagger blade grass shrimp (*Palaemonetes pugio*). *Environ. Toxicol. Chem.* **2017**, *36*, 3074–3080. [CrossRef] [PubMed]
6. Ziajahromi, S.; Kumar, A.; Neale, P.A.; Leusch, F.D.L. Impact of microplastic beads and fibers on waterflea (*Ceriodaphnia dubia*) survival, growth, and reproduction: Implications of single and mixture exposures. *Environ. Sci. Technol.* **2017**, *51*, 13397–13406. [CrossRef] [PubMed]

proceedings

Abstract

Synthesis and Antibacterial Properties of Lignin-Based Quaternary Ammonium and Phosphonium Salts †

Mahendra K. Mohan [1], Harleen Kaur [2], Ella Duvanova [1], Merilin Rosenberg [2], Marcos Dahlem [3], Angela Ivask [2], Jean-Manuel Raimundo [3], Tiit Lukk [1] and Yevgen Karpichev [1,*]

1 Department of Chemistry and Biotechnology, Tallinn University of Technology (TalTech), 15 Akadeemia tee, 12618 Tallinn, Estonia; mahendra.mohan@taltech.ee (M.K.M.); ella.ivantsova@gmail.com (E.D.); tiit.lukk@taltech.ee (T.L.)
2 Institute of Molecular and Cell Biology, University of Tartu, 23 Riia St., 51010 Tartu, Estonia; harleen.kaur@ut.ee (H.K.); merilin.rosenberg@ut.ee (M.R.); angela.ivask@ut.ee (A.I.)
3 Département Matériaux Polytech Marseille, Aix Marseille University, CINaM UMR CNRS 7325, 163 Avenue de Luminy, CEDEX 09, 13288 Marseille, France; marcosdahlem@gmail.com (M.D.); jean-manuel.raimundo@univ-amu.fr (J.-M.R.)
* Correspondence: yevgen.karpichev@taltech.ee; Tel.: +372-620-4381
† Presented at the International Conference EcoBalt 2023 "Chemicals & Environment", Tallinn, Estonia, 9–11 October 2023.

Keywords: lignin; quaternisation; antibacterial

Lignin, a naturally occurring aromatic polymer, possesses a range of biological functions. Notably, plants have harnessed lignin's properties as a defense mechanism against invading pathogenic microbes. Consequently, the utilization of isolated lignin as an ecofriendly antimicrobial agent holds great promise for enhancing the value of lignin. Furthermore, given lignin's green and sustainable origin through plant photosynthesis, its integration into the antimicrobial industry has the potential to reduce carbon emissions. Numerous studies have explored the utilization of lignin for the development of antimicrobial agents tailored to various applications. However, lignin is a highly heterogeneous polymer, characterized by variations in monomer composition, linkages, molecular weight, and functional groups. Consequently, the relationship between lignin's structure and properties, as well as its mechanism of action as an antimicrobial agent, remains unclear [1]. To address these gaps in knowledge, we conducted a study in which we prepared forty-two quaternary ammonium/phosphonium organosolv lignin samples from aspen, pine, and barley straw (representing hardwood, softwood, and grass sources, correspondingly) using a versatile intermediate, known as chloromethylated lignin, developed recently [2]. These samples were then assessed for their potential antibacterial properties against clinical isolates of Gram-positive (*Staphylococcus aureus*, *MRSA*) and Gram-negative (*Klebsiella pneumonia*) pathogenic bacterial strains. Our findings suggest that the antibacterial activity of these lignin samples exhibits an increase with the length of the hydrophobic chain, up to C14, after which it begins to decline. Additionally, ongoing research is being conducted to investigate the antimicrobial activity of ammonium and phosphonium surfactant materials, both individually and in combination. In this research, chemical modification has significantly boosted the antimicrobial efficacy of lignin, and incorporating extra chemical structures, e.g., cationic functional groups through chemical modification, presents a viable strategy to enhance the future potential of lignin's antimicrobial activity.

Author Contributions: Conceptualization, methodology, investigation, original draft preparation: M.K.M. and Y.K.; investigation: H.K., M.D. and E.D.; validation: M.R.; resources, supervision, project administration, and funding acquisition: A.I., T.L., J.-M.R. and Y.K. All authors have read and agreed to the published version of the manuscript.

Citation: Mohan, M.K.; Kaur, H.; Duvanova, E.; Rosenberg, M.; Dahlem, M.; Ivask, A.; Raimundo, J.-M.; Lukk, T.; Karpichev, Y. Synthesis and Antibacterial Properties of Lignin-Based Quaternary Ammonium and Phosphonium Salts. *Proceedings* **2023**, *92*, 58. https://doi.org/10.3390/proceedings2023092058

Academic Editors: Monika Mortimer, Anne Kahru, Ivo Leito, Riin Rebane and Villem Aruoja

Published: 30 November 2023

Funding: This research was funded by Estonian Research Council grants RESTA11 (for M.K.M., T.L., Y.K.), COVSG5 (for M.K.M., E.D., Y.K.) and PRG1496 (A.I., M.R., H.K). Funding was also obtained from ERDF Dora Plus programme (for E.D.), and PARROT French-Estonian science and technology cooperation programme (for M.K.M., M.D., J.-M.R, Y.K.).

Institutional Review Board Statement: Not applicable.

Informed Consent Statement: Not applicable.

Data Availability Statement: The data presented in this study are available on request from the corresponding author. The data are not publicly available due to manuscript preparation.

Conflicts of Interest: The authors declare no conflict of interest.

References

1. Chen, M.; Li, Y.; Liu, H.; Zhang, D.; Shi, Q.S.; Zhong, X.Q.; Guo, Y.; Xie, X.B. High value valorization of lignin as environmental benign antimicrobial. *Mater. Today Bio* **2023**, *18*, 100520. [CrossRef] [PubMed]
2. Mohan, M.K.; Silenko, O.; Krasnou, I.; Volobujeva, O.; Kulp, M.; Ošeka, M.; Lukk, T.; Karpichev, Y. Chloromethylated Lignin as a Precursor of Sustainable Catalytic Material for Cross-Coupling and Click Reactions. Available online: https://chemrxiv.org/engage/chemrxiv/article-details/653256a287198ede07e00d40 (accessed on 30 October 2023).

proceedings

MDPI

Abstract

Application of Fly Ash of Lignite Combustion in Air and Water Purification †

Dmitri Nikitin [1], Juri Bolobajev [1], Marina Krichevskaya [1], Lukas Pilar [2], Monika Vitvarova [3], Sergei Preis [1] and Niina Dulova [1,*]

1 Department of Materials and Environmental Technology, Tallinn University of Technology, Ehitajate tee 5, 19086 Tallinn, Estonia; dmitri.nikitin@taltech.ee (D.N.); juri.bolobajev@taltech.ee (J.B.); marina.kritsevskaja@taltech.ee (M.K.); sergei.preis@taltech.ee (S.P.)

2 Energy Department, Faculty of Mechanical Engineering, Czech Technical University in Prague, Technická 4, 166 07 Prague, Czech Republic; lukas.pilar@fs.cvut.cz

3 Department of Sustainability and Product Ecology, University of Chemistry and Technology Prague, Technická 5, 160 00 Prague, Czech Republic; monika.vitvarova@vscht.cz

* Correspondence: niina.dulova@taltech.ee

† Presented at the International Conference EcoBalt 2023 "Chemicals & Environment", Tallinn, Estonia, 9–11 October 2023.

Keywords: heterogenous catalyst; adsorption; advanced oxidation processes; fly ash; sustainable waste management

The study was aimed at evaluating the catalytic and photocatalytic properties of lignite fly ash samples, S1 and S2, from the Pilsen Power Station (Teplárna Plzeň) collected using electrostatic precipitators with relatively high contents of Fe_2O_3 (6.0–7.4%) and TiO_2 (4.6–4.8%). Iron oxides are used often as heterogeneous Fenton(-like) reaction catalysts in certain compositions, as they have iron oxides and oxo-hydroxides attached to catalyst supports with developed contact surfaces, such as zeolites [1,2]. TiO_2 is the most studied photocatalyst for the photocatalytic oxidation of gaseous volatile organic compounds (VOCs) [3]. Experimental tests investigating the catalytic oxidation of the textile dye Acid Orange 7 in aqueous solutions, using the heterogenous Fenton-like system (H_2O_2/fly ash) and the photocatalytic oxidation of acetone vapors in the UV-A/fly ash system, were carried out. In water treatment trials, adsorption and Fenton-like experiments were carried out in parallel, under similar treatment conditions. The adsorption of the VOC was characterized by its concentration in the reactor's effluent growing with the adsorbent saturation with the acetone vapors. Once the studied sample of fly ash had accumulated a certain amount of acetone, the UV-A light was switched on to start the photocatalytic oxidation reaction. The results showed moderate catalytic and negligible photocatalytic properties of S1 and S2 in the studied systems, although they exhibited certain adsorption properties. Surprisingly, S2 showed a noticeably stronger catalytic ability in the Fenton-like system than S1, despite having an almost 2.6 times lower surface area at a similar chemical composition. The fly ashes were also used for zeolite synthesis [4]. They were subsequently tested in ion exchange with respect to ammonium cations and showed abilities close to those of a commercial zeolite specimen.

Citation: Nikitin, D.; Bolobajev, J.; Krichevskaya, M.; Pilar, L.; Vitvarova, M.; Preis, S.; Dulova, N. Application of Fly Ash of Lignite Combustion in Air and Water Purification. *Proceedings* **2023**, *92*, 32. https://doi.org/10.3390/proceedings2023092032

Academic Editors: Monika Mortimer, Anne Kahru, Ivo Leito, Riin Rebane and Villem Aruoja

Published: 22 November 2023

Author Contributions: Conceptualization, J.B., N.D., and M.K.; methodology, J.B., N.D., and M.K.; formal analysis, J.B., N.D., L.P., and M.K.; investigation, D.N., J.B., N.D., and M.K.; resources, L.P., M.V., and S.P.; writing—original draft preparation, N.D. and M.K.; writing—review and editing, N.D., and S.P.; visualization, J.B., N.D., and M.K.; supervision, N.D. and S.P.; funding acquisition, M.V., and S.P. All authors have read and agreed to the published version of the manuscript.

Funding: This work has received funding from the European Union's Horizon 2020 Research and Innovation Programme, ERA-NET Confund on Raw Materials (ERA-MIN3) via the project ABtomat (JTC-2021_131).

Data Availability Statement: Data supporting the results presented can be provided upon request to the corresponding author.

Conflicts of Interest: The authors declare no conflict of interest. The funders had no role in the design of the study; in the collection, analyses, or interpretation of data; in the writing of the manuscript; or in the decision to publish the results.

References

1. Dulova, N.; Trapido, M.; Dulov, A. Catalytic degradation of picric acid by heterogeneous Fenton-based processes. *Environ. Technol.* **2011**, *32*, 439–446. [CrossRef]
2. Ioffe, M.; Kundu, S.; Perez-Lapid, N.; Radian, A. Heterogeneous Fenton catalyst based on clay decorated with nano-sized amorphous iron oxides prevents oxidant scavenging through surface complexation. *Chem. Eng. J.* **2022**, *433*, 134609. [CrossRef]
3. Jang, W.-S.; Pham, V.N.; Yang, S.-H.; Baik, J.; Lee, H.; Kim, Y.-M. Enhanced selective photocatalytic oxidation of a bio-derived platform chemical with vacancy-induced core-shell anatase TiO_2 nanoparticles. *Appl. Cat. B Environ.* **2023**, *322*, 122140. [CrossRef]
4. Imbert, F.E.; Moreno, C.; Montero, A.; Fontal, B.; Lujano, J. Venezuelan natural alumosilicates as a feedstock in the synthesis of zeolite A. *Zeolites* **1994**, *14*, 374–378. [CrossRef]

proceedings

MDPI

Abstract

Crystal and Magnetic Structure Transitions in Doped Lu- and Fe-Based Perovskite Oxides †

Andrius Pakalniškis [1,*], Gediminas Niaura [2], Dmitry Karpinsky [3], Guillaume Rogez [4], Pierre Rabu [4], Shih-Wen Chen [5,6], Thomas Chung-Kuang Yang [5,6], Ramūnas Skaudžius [1] and Aivaras Kareiva [1]

1 Institute of Chemistry, Vilnius University, 01513 Vilnius, Lithuania
2 Department of Organic Chemistry, Center for Physical Sciences and Technology (FTMC), 10257 Vilnius, Lithuania
3 Namangan Engineering-Construction Institute, Dustlik Avenue 4, Namangan 160100, Uzbekistan
4 IPCMS (UMR 7504 CNRS), University of Strasbourg, 67000 Strasbourg, France
5 Precision Analysis and Materials Research Center, National Taipei University of Technology, Taipei 10608, Taiwan
6 Department of Chemical Engineering and Biotechnology, National Taipei University of Technology, Taipei 10608, Taiwan
* Correspondence: andrius.pakalniskis@chgf.vu.lt
† Presented at the International Conference EcoBalt 2023 "Chemicals & Environment", Tallinn, Estonia, 9–11 October 2023.

Keywords: multiferroic; perovskite; magnetization

The possibility of controlling physical properties via chemical doping is particularly important when considering the formation of both electrical and magnetic orderings in the same compounds, which are commonly referred as multiferroics [1]. However, for the most part, due to the conflicting nature of these properties, the coupling between the electrical and magnetic properties is relatively weak. Recently, a new class of hexagonal rare earth ferrite perovskite compounds has been found to exhibit multiferroic ordering, with a mechanism and structure similar to that of hexagonal manganites, making them a new avenue for potential research. This new family of room temperature multiferroic compounds is based on $LuFeO_3$ with a hexagonal structure (space group $P6_3cm$). It has been discovered that $LuFeO_3$ in its hexagonal state has both ferroelectric and weak ferromagnetic orderings [2,3]. As such, in this work, we adapted an ethylene glycol based sol-gel synthesis procedure for the preparation of bulk Sc doped hexagonal lutetium ferrite powders. The crystal structure was investigated using XRD and Raman spectroscopy at room temperature. The temperature-driven crystal structure transitions were analyzed by means of in situ high-temperature XRD while the magnetic transitions were investigated by means of low-temperature magnetization measurements. The obtained results revealed that at room temperature, the polar hexagonal phase can be stabilized in a quite narrow doping range that depends on the sintering temperature of the samples [4]. Additionally, samples with a higher Sc doping content showed a lower phase transition temperature from ferroelectric to paraelectric phases. While Sc itself is not magnetic, the Sc doping caused substantial changes to the magnetization of the samples as well. Overall, the results indicate that $LuFeO_3$ can be successfully synthesized by means of the ethylene glycol based sol-gel procedure. The desired phase composition, magnetic and electric properties can be optimized by means of the doping content and the sintering temperature.

Citation: Pakalniškis, A.; Niaura, G.; Karpinsky, D.; Rogez, G.; Rabu, P.; Chen, S.-W.; Yang, T.C.-K.; Skaudžius, R.; Kareiva, A. Crystal and Magnetic Structure Transitions in Doped Lu- and Fe-Based Perovskite Oxides. *Proceedings* **2023**, *92*, 27. https://doi.org/10.3390/proceedings2023092027

Published: 22 November 2023

Author Contributions: Conceptualization, A.K., R.S. and D.K.; methodology, A.P. and S.-W.C.; formal analysis, A.P., D.K., G.R. and G.N.; investigation, A.P., G.R., S.-W.C. and G.N.; resources, P.R., A.K. and T.C.-K.Y.; data curation, R.S. and P.R.; writing—original draft preparation, G.R., A.P. and D.K.; writing—review and editing, D.K. and A.P.; visualization, A.P, S.-W.C., D.K. and G.R.; supervision, R.S., G.R. and P.R.; project administration, A.K.; funding acquisition, A.K., T.C.-K.Y. and P.R. All authors have read and agreed to the published version of the manuscript.

Funding: This project has received funding from the European Union's Horizon 2020 research and innovation program under the Marie Skłodowska-Curie grant agreement No 778070—TransFerr—H2020-MSCA-RISE-2017.

Institutional Review Board Statement: Not applicable.

Informed Consent Statement: Not applicable.

Data Availability Statement: Data Availability Statements are available in section.

Conflicts of Interest: The authors declare no conflict of interest.

References

1. Fiebig, M.; Lottermoser, T.; Meier, D.; Trassin, M. The evolution of multiferroics. *Nat. Rev. Mater.* **2016**, *1*, 16046. [CrossRef]
2. Wang, W.; Zhao, J.; Wang, W.; Gai, Z.; Balke, N.; Chi, M.; Lee, H.N.; Tian, W.; Zhu, L.; Cheng, X.; et al. Room-temperature multiferroic hexagonal $LuFeO_3$ films. *Phys. Rev. Lett.* **2013**, *110*, 237601. [CrossRef] [PubMed]
3. Cao, S.; Zhang, X.; Paudel, T.R.; Sinha, K.; Wang, X.; Jiang, X.; Wang, W.; Brutsche, S.; Wang, J.; Ryan, P.J.; et al. On the structural origin of the single-ion magnetic anisotropy in $LuFeO_3$. *J. Phys. Condens. Matter* **2016**, *28*, 156001. [CrossRef] [PubMed]
4. Pakalniškis, A.; Alikin, D.O.; Turygin, A.P.; Zhaludkevich, A.L.; Silibin, M.V.; Zhaludkevich, D.V.; Niaura, G.; Zarkov, A.; Skaudžius, R.; Karpinsky, D.V.; et al. Crystal Structure and Concentration-Driven Phase Transitions in $Lu_{(1-x)}Sc_xFeO_3$ ($0 \leq x \leq 1$) Prepared by the Sol–Gel Method. *Materials* **2022**, *15*, 1048. [CrossRef] [PubMed]

proceedings

Abstract

Influence of Black Alder Bark Extractives as Integral Building Blocks on the Susceptibility to Biodegradation of Resilient Polyether Polyurethanes †

Jevgenija Ponomarenko *, Matiss Pals, Oskars Bikovens and Alexandr Arshanitsa

Laboratory of Lignin Chemistry, Latvian State Institute of Wood Chemistry, 1006 Riga, Latvia; matiss.pals@kki.lv (M.P.); oskars.bikovens@kki.lv (O.B.); alexandr.arsanica@kki.lv (A.A.)

* Correspondence: jevgenija.ponomarenko@kki.lv

† Presented at the International Conference EcoBalt 2023 "Chemicals & Environment", Tallinn, Estonia, 9–11 October 2023.

Keywords: polyurethanes; bio-based; bark extractives; polyols; biological degradation

In response to the increasing demand for environmentally friendly materials and the necessity of effective end-of-life management [1], this study investigates the susceptibility of innovative polyurethanes (PUs) that incorporate plant extractives to biodegradation. The PUs were synthesized using polymeric methylene diphenyl diisocyanate, polyol PEG 400, and tetrahydrofuran as a solvent, with NCO/OH ratios of 1.0, 0.8, and 0.5 [2]. Partial or complete substitution of the conventional PEG 400 polyol with black alder bark extracts as bio-polyols was performed. Degradation tests were conducted in sewage water and compost-enriched soil [3,4]. The susceptibility to biodegradation was assessed through weight loss measurements, FTIR spectroscopy, and biological oxygen demand (BOD) analysis [5–7]. After 2 months, conventional PUs exhibited weight losses of 9.6% in soil and 12.4% in water. Complete replacement of PEG 400 with bark-sourced polyol increased weight loss to 15.6% in soil and 15.7% in water. The reference biodegradable polylactic acid (PLA) film displayed weight losses of 35.6% in soil and 15.0% in water. BOD measurements indicated that PUs, particularly those containing bio-based building blocks, supported microbial metabolism, corroborating previous findings [8]. The incorporation of bark extractives significantly enhanced the susceptibility of PUs to biodegradation, resulting in a reduction of up to 45% in the intensity of FTIR spectral peaks associated with the urethane functional group (-NHC(=O)-O-), compared to a maximum reduction of 11% for conventional PUs. Moreover, it displayed increased peak intensities for O-H and N-H stretches after biodegradation, exhibiting a notable negative correlation with changes in peak intensities for N-CO-O, C-O-C, and C-N vibrations following PU biodegradation. This indicates the formation of hydroxyl and amino groups resulting from the hydrolysis of various chemical bonds within the polyurethane network, facilitated by the integration of bark extractives. This work emphasizes the potential of bark-derived building blocks in the design of PU materials suitable for biological recycling while recognizing the need for further investigation into the optimal agents and conditions for the biological conversion of the latter.

Citation: Ponomarenko, J.; Pals, M.; Bikovens, O.; Arshanitsa, A. Influence of Black Alder Bark Extractives as Integral Building Blocks on the Susceptibility to Biodegradation of Resilient Polyether Polyurethanes. *Proceedings* **2023**, *92*, 57. https://doi.org/10.3390/proceedings2023092057

Academic Editors: Monika Mortimer, Anne Kahru, Ivo Leito, Riin Rebane and Villem Aruoja

Published: 29 November 2023

Author Contributions: Conceptualization, J.P. and A.A.; methodology, A.A.; validation, J.P., A.A., M.P. and O.B.; formal analysis, J.P., M.P. and O.B.; investigation, J.P.; data curation, J.P.; writing—original draft preparation, J.P.; writing—review and editing, J.P., A.A., M.P. and O.B.; visualization, J.P., A.A., M.P. and O.B.; supervision, A.A.; project administration, A.A.; funding acquisition, A.A. All authors have read and agreed to the published version of the manuscript.

Funding: This research was funded by the Latvian Budget—the State Research Programme Nr. lzp-2021/1-0207.

Institutional Review Board Statement: Not applicable.

Informed Consent Statement: Not applicable.

Data Availability Statement: Data will be made available on request.

Conflicts of Interest: The authors declare no conflict of interest.

References

1. Magnin, A.; Pollet, E.; Phalip, V.; Avérous, L. Evaluation of Biological Degradation of Polyurethanes. *Biotechnol. Adv.* **2020**, *39*, 107457. [CrossRef] [PubMed]
2. Arshanitsa, A.; Krumina, L.; Telysheva, G.; Dizhbite, T. Exploring the Application Potential of Incompletely Soluble Organosolv Lignin as a Macromonomer for Polyurethane Synthesis. *Ind. Crop. Prod.* **2016**, *92*, 1–12. [CrossRef]
3. Pischedda, A.; Tosin, M.; Degli-Innocenti, F. Biodegradation of Plastics in Soil: The Effect of Temperature. *Polym. Degrad. Stab.* **2019**, *170*, 109017. [CrossRef]
4. Nabeoka, R.; Suzuki, H.; Akasaka, Y.; Ando, N.; Yoshida, T. Evaluating the Ready Biodegradability of Biodegradable Plastics. *Environ. Toxicol. Chem.* **2021**, *40*, 2443–2449. [CrossRef] [PubMed]
5. Kim, J.-H.; Choi, S.H.; Park, M.G.; Park, D.H.; Son, K.-H.; Park, H.-Y. Biodegradation of Polyurethane by Japanese Carpenter Bee Gut-Associated Symbionts Xanthomonas Sp. HY-71, and Its Potential Application on Bioconversion. *Environ. Technol. Innov.* **2022**, *28*, 102822. [CrossRef]
6. Gogoi, S.; Karak, N. Biobased Biodegradable Waterborne Hyperbranched Polyurethane as an Ecofriendly Sustainable Material. *ACS Sustain. Chem. Eng.* **2014**, *2*, 2730–2738. [CrossRef]
7. Zuliani, A.; Rapisarda, M.; Chelazzi, D.; Baglioni, P.; Rizzarelli, P. Synthesis, Characterization, and Soil Burial Degradation of Biobased Polyurethanes. *Polymers* **2022**, *14*, 4948. [CrossRef] [PubMed]
8. Gunawan, N.R.; Tessman, M.; Zhen, D.; Johnson, L.; Evans, P.; Clements, S.M.; Pomeroy, R.S.; Burkart, M.D.; Simkovsky, R.; Mayfield, S.P. Biodegradation of Renewable Polyurethane Foams in Marine Environments Occurs through Depolymerization by Marine Microorganisms. *Sci. Total Environ.* **2022**, *850*, 158761. [CrossRef] [PubMed]

proceedings

MDPI

Abstract

Hazard Evaluation of Novel Plasticizer, Di(2-Propylheptyl) Phthalate, to Aquatic Ecosystems †

Anastasija Petrova [1,2], Aljona Lukjanova [1,*], Irina Blinova [1], Heiki Vija [1] and Margit Heinlaan [1]

1 Laboratory of Environmental Toxicology, National Institute of Chemical Physics and Biophysics, Akadeemia Tee 23, 12618 Tallinn, Estonia; nastja37@tlu.ee (A.P.); irina.blinova@kbfi.ee (I.B.); heiki.vija@kbfi.ee (H.V.); margit.heinlaan@kbfi.ee (M.H.)

2 School and Natural Sciences and Health, Tallinn University, Uus Sadama 5, 10120 Tallinn, Estonia

* Correspondence: aljona.lukjanova@kbfi.ee

† Presented at the International Conference EcoBalt 2023 "Chemicals & Environment", Tallinn, Estonia, 9–11 October 2023.

Keywords: microcrustacean; ecotoxicity; DPHP; DEHP; *Daphnia magna*

Citation: Petrova, A.; Lukjanova, A.; Blinova, I.; Vija, H.; Heinlaan, M. Hazard Evaluation of Novel Plasticizer, Di(2-Propylheptyl) Phthalate, to Aquatic Ecosystems. *Proceedings* **2023**, *92*, 45. https://doi.org/10.3390/proceedings2023092045

Academic Editors: Monika Mortimer, Anne Kahru, Ivo Leito, Riin Rebane and Villem Aruoja

Published: 24 November 2023

Plastics differ in chemical composition due to numerous chemicals (additives) used to produce plastic for specific applications. Plastic additives may enter the environment not only from the plastic waste during its weathering but also at all stages of the plastic products' life cycle [1]. Plastic additives can be hazardous to living organisms and play a significant role in the adverse effects of plastic contamination [2,3]. However, there are knowledge gaps concerning the ecotoxicity of plastic additives [4]. Plasticizers are one group of plastic additives of which phthalates are the most used, but they are also known for their toxic potential. As the use of phthalate plasticizers has been increasingly regulated since the beginning of the 21st century, alternative plasticizers have appeared in their stead. The current study was conducted to obtain new experimental data on the potential ecotoxicity of high molecular weight phthalate plasticizer DPHP (Di(2-propylheptyl) phthalate) in comparison to DEHP (Di-2-ethylhexyl phthalate), which was a formerly dominant but now restricted [5] low molecular weight plasticizer. For hazard evaluation, long-term (21 day) effects were studied in life cycle tests with *Daphnia magna*, which is a representative of microcrustacea and important link of the freshwater food web. *D. magna* was exposed to the plasticizers via spiked sediment (sand).The tests were conducted in lake water to increase the environmental relevance of the hazard data. Alarmingly, preliminary results showed that DPHP may be more hazardous to aquatic organisms than DEHP. The mortality of the organisms, exposed to DPHP plasticizer, was higher than with DEHP. *D. magna* reproduction (the average number of offspring/organism) was lower than in the unexposed control and comparable for both DPHP and DEHP. However, since fewer offspring were also recorded for the used solvent (ethyl acetate) control, the affected reproduction was not necessarily attributable to the plasticizers.

Author Contributions: Conceptualization, M.H., I.B. and A.L.; methodology, I.B., A.L. and H.V.; investigation, A.L., I.B., A.P. and H.V.; resources, M.H.; data curation, A.L., I.B. and H.V.; writing—original draft preparation, I.B. and A.L.; writing—review and editing, I.B., A.L. and M.H.; project administration M.H.; funding acquisition, M.H. All authors have read and agreed to the published version of the manuscript.

Funding: This research was funded by the Estonian Research Council grant PRG1427.

Institutional Review Board Statement: Not applicable.

Informed Consent Statement: Not applicable.

Data Availability Statement: Not applicable.

Acknowledgments: The authors wish to thank AS Tallinna Vesi for cooperation (access to Ülemiste lake water and chemical analyses data of Ülemiste lake water).

Conflicts of Interest: The authors declare no conflict of interest. The funders had no role in the design of the study; in the collection, analyses, or interpretation of data; in the writing of the manuscript; or in the decision to publish the results.

References

1. Hahladakis, J.; Velis, C.; Weber, R.; Iacovidou, E.; Purnell, P. An overviez of chemical additives present in plastics: Migration, release, fate and environmental impact during their use, disposal and recycling. *J. Hazard. Mater.* **2018**, *344*, 179–199. [CrossRef] [PubMed]
2. Beiras, R.; Verdejo, E.; Campoy-López, P.; Vidal-Liñán, L. Aquatic toxicity of chemically defined microplastics can be explained by functional additives. *J. Hazard. Mater.* **2021**, *406*, 124338. [CrossRef] [PubMed]
3. Gunaalan, K.; Fabbri, E.; Capolupo, M. The hidden threat of plastic leachates: A critical review on their impacts on aquatic organisms. *Water Res.* **2020**, *184*, 116170. [CrossRef] [PubMed]
4. Barrick, A.; Champeau, O.; Chatel, A.; Manier, N.; Northcott, G.; Tremblay, L.A. Plastic additives: Challenges in ecotox hazard assessment. *PeerJ* **2021**, *9*, e11300. [CrossRef] [PubMed]
5. ECHA (European Chemicals Agency). Substances Restricted under Reach—Annex XVII Entry 51. 2022. Available online: https://www.echa.europa.eu/substances-restricted-under-reach (accessed on 20 September 2023).

proceedings

MDPI

Abstract

Buckwheat Husks, Ash and Biomass for Sustainable Plant Fertilization and Soil Improvement †

Odeta Pocienė * and Rasa Šlinkšienė

Department of Physical and Inorganic Chemistry, Kaunas University of Technology, LT-50254 Kaunas, Lithuania; rasa.slinksiene@ktu.lt
* Correspondence: odeta.pociene@ktu.lt

† Presented at the International Conference EcoBalt 2023 "Chemicals & Environment", Tallinn, Estonia, 9–11 October 2023.

Keywords: buckwheat; organic fertilizers; granulation

Nowadays, it is difficult to imagine agriculture without the use of fertilizers, because plants cannot absorb the required amount of nutrients from the soil. The only way to provide plants with all the nutrients they need is to fertilize them. Depending on the properties of the soil, the type of plant, climatic conditions, etc., different amounts of nutrients are required for effective plant growth [1]. However, the intensive and unbalanced use of concentrated mineral fertilizers reduces the reserves of fertilizer raw materials, exhausts the soil and harms the ecosystem. Meanwhile, in some industries (food, energy, etc.), increasing amounts of production by-products are generated, which could be processed into high-quality organic fertilizers. The decomposing matter from organic fertilizers breaks down naturally and would provide nutrient and minerals to the soil [2,3]. One such material is waste from the buckwheat groats industry: uncleaned buckwheat biomass (UBM), buckwheat husks (BH) and buckwheat husk ash (BHA). These wastes contain many different nutrients that plants need, so it becomes possible to use them as fertilizers. It is difficult to directly use biofuel ash, buckwheat hulls or biomass for soil fertilization because, due to the improper shape, non-uniformity and high dustiness of their particles, the maximum fertilization efficiency is not achieved, and it is necessary to granulate them. A drum granulator was used to produce a mixture of raw materials containing various wastes: BH, BHA and UBM. Depending on the composition of the starting materials, it was possible to obtain up to 60% of a productive fraction. The moisture content of such granules varied between 2 and 10% and they had a relatively low bulk density (between 430–480 kg/m^3). The pH values of the 10% solution of the produced granules ranged from 9.7 to 12.0, which indicates that the fertilizer can act as a lime agent; it is recommended to fertilize acidic soils. It should also be mentioned that the obtained pellets are quite weak and plastic. Granular fertilizers are non-hygroscopic and retain their shape when stored at 21–23 °C and 70–75% humidity. In summary, it can be said that buckwheat groat waste can be used in the production of environmentally friendly fertilizers in order to reduce environmental pollution, slow down soil degradation and increase the amount of soil organic matter.

Citation: Pocienė, O.; Šlinkšienė, R. Buckwheat Husks, Ash and Biomass for Sustainable Plant Fertilization and Soil Improvement. *Proceedings* **2023**, *92*, 53. https://doi.org/10.3390/proceedings2023092053

Academic Editors: Monika Mortimer, Anne Kahru, Ivo Leito, Riin Rebane and Villem Aruoja

Published: 27 November 2023

Author Contributions: Conceptualization, R.Š. and O.P.; methodology, R.Š. and O.P.; software, O.P.; formal analysis, O.P.; investigation, O.P.; resources, O.P.; data curation, R.Š.; writing—original draft preparation, O.P.; writing—review and editing, R.Š.; supervision, R.Š. All authors have read and agreed to the published version of the manuscript.

Funding: This research received no external funding.

Institutional Review Board Statement: Not applicable.

Informed Consent Statement: Not applicable.

Data Availability Statement: Data sharing is not applicable.

Conflicts of Interest: The authors declare no conflict of interest.

References

1. Javed, A.; Ali, E.; Afzal, K.B.; Osman, A.; Riaz, S. Soil Fertility: Factors Affecting Soil Fertility, and Biodiversity Responsible for Soil Fertility. *Int. J. Plant Anim. Environ. Sci.* **2022**, *12*, 21–33. [CrossRef]
2. Assefa, S.; Tadesse, S. The Principal Role of Organic Fertilizer on Soil Properties and Agricultural Productivity—A Review. *Agric. Res. Technol.* **2019**, *22*, 46–50.
3. Ye, L.; Zhao, X.; Bao, E.; Li, J.; Zou, Z.; Cao, K. Bio-organic fertilizer with reduced rates of chemical fertilization improves soil fertility and enhances tomato yield and quality. *Sci. Rep.* **2020**, *10*, 177. [CrossRef] [PubMed]

proceedings

Abstract

Effect of MgO Templating on the Synthesis and Properties of Dissolved Lignin-Based Hard Carbon for Na-Ion Battery Applications †

Angelo Robiños [1], Hao Zhang [2], Zekra Mousavi [1], Jan-Henrik Smått [1], Tor Laurén [1], Leena Hupa [1], Chunlin Xu [2] and Johan Bobacka [1,*]

[1] Laboratory of Molecular Science and Engineering, Faculty of Science and Engineering, Åbo Akademi University, Henriksgatan 2, 20500 Åbo-Turku, Finland; angelo.robinos@abo.fi (A.R.); zekra.mousavi@abo.fi (Z.M.); jan-henrik.smatt@abo.fi (J.-H.S.); tor.lauren@abo.fi (T.L.); leena.hupa@abo.fi (L.H.)

[2] Laboratory of Natural Materials Technology, Faculty of Science and Engineering, Åbo Akademi University, Henriksgatan 2, 20500 Åbo-Turku, Finland; hao.zhang@abo.fi (H.Z.); chunlin.xu@abo.fi (C.X.)

* Correspondence: johan.bobacka@abo.fi; Tel.: +358-46-9200208

† Presented at the International Conference EcoBalt 2023 "Chemicals & Environment", Tallinn, Estonia, 9–11 October 2023.

Keywords: lignin; hard carbon; Na-ion battery; porous material; template synthesis

The increasing use of wind and solar energy creates an enormous need for intermittent storage of electrical energy in batteries. Today, Li-ion batteries are the state of the art in mobile applications, such as electric vehicles. However, due to the limited sources of lithium, there is a growing need to replace Li-ion batteries with more sustainable alternatives, such as Na-ion batteries. This development involves the replacement of graphite with alternative anode materials, such as hard carbon. Lignin, a naturally abundant biopolymer, has shown promising potential as a carbon precursor for electrical energy storage applications, particularly in the synthesis of hard carbon anodes for Na-ion batteries [1]. In this study, we investigate the synthesis of lignin-based hard carbon using a MgO template technique, where lignin is dissolved using NaOH. The effect of the synthesis process on the morphology, porous structure, and electrochemical properties of the resulting hard carbon material is investigated. The synthesis process involves the carbonization of freeze-dried solutions containing dissolved lignin and magnesium gluconate [2]. By subjecting the mixture to preheat treatment at 600 °C, nano-sized domains of Mg and Na crystals form within the carbon matrix. Acid leaching of the resulting particles is subsequently carried out, followed by high-temperature post-heat treatment at 1100–1500 °C. These lead to the formation of a hierarchical porous hard carbon structure for Na-ion battery applications. The findings from this research have the potential to contribute to the development of sustainable and high-performance energy storage systems.

Citation: Robiños, A.; Zhang, H.; Mousavi, Z.; Smått, J.-H.; Laurén, T.; Hupa, L.; Xu, C.; Bobacka, J. Effect of MgO Templating on the Synthesis and Properties of Dissolved Lignin-Based Hard Carbon for Na-Ion Battery Applications. *Proceedings* **2023**, *92*, 16. https://doi.org/10.3390/proceedings2023092016

Academic Editors: Monika Morti-mer, Anne Kahru, Ivo Leito, Riin Rebane and Villem Aruoja

Published: 21 November 2023

Author Contributions: Conceptualization, A.R., H.Z., C.X. and J.B.; methodology, A.R., H.Z. and Z.M.; formal analysis, J.-H.S.; investigation, A.R., H.Z. and Z.M.; supervision, J.B., C.X. and L.H.; project administration, J.B., C.X. and T.L.; funding acquisition, J.B. and C.X. All authors have read and agreed to the published version of the manuscript.

Funding: This research was funded by the Jane and Aatos Erkko Foundation, Finland (Grant number 220015).

Institutional Review Board Statement: Not applicable.

Informed Consent Statement: Not applicable.

Data Availability Statement: The data presented in this study are available upon request from the corresponding author.

Acknowledgments: The authors gratefully acknowledge CH-bioforce, Finland, for generously providing the birch lignin used in this study.

Conflicts of Interest: The authors declare no conflict of interest. The funders had no role in the design of the study; in the collection, analyses, or interpretation of data; in the writing of the manuscript; or in the decision to publish the results.

References

1. Matei Ghimbeu, C.; Zhang, B.; Martinez de Yuso, A.; Réty, B.; Tarascon, J.-M. Valorizing Low Cost and Renewable Lignin as Hard Carbon for Na-Ion Batteries: Impact of Lignin Grade. *Carbon* **2019**, *153*, 634–647. [CrossRef]
2. Kamiyama, A.; Kubota, K.; Igarashi, D.; Youn, Y.; Tateyama, Y.; Ando, H.; Gotoh, K.; Komaba, S. MgO-Template Synthesis of Extremely High Capacity Hard Carbon for Na-Ion Battery. *Angew. Chem. Int. Ed.* **2021**, *60*, 5114–5120. [CrossRef]

proceedings

MDPI

Abstract

Solid Solution Formation in Xanthone–Thioxanthone Binary System: Experimental Investigation †

Kristaps Saršūns *, Agris Bērziņš and Toms Rekis

Department of Physical Chemistry, Faculty of Chemistry, University of Latvia, LV-1004 Riga, Latvia; agris.berzins@lu.lv (A.B.); toms.rekis@lu.lv (T.R.)

* Correspondence: kristaps.sarsuns@lu.lv; Tel.: +371-22354160

† Presented at the International Conference EcoBalt 2023 "Chemicals & Environment", Tallinn, Estonia, 9–11 October 2023.

Keywords: solid solution; thermal analysis; binary phase diagram

Solid solutions are crystalline phases consisting of at least two components in a freely variable composition within certain limits. Depending on the miscibility of the components, solid solutions are divided into two types, i.e., solid solutions of unlimited solubility of components and limited solubility of components [1]. Solid solutions among inorganic substances are widely studied and well-known solid phases for which the structure and properties of the material are dependent on the component ratio. Such phases and the change of properties they provide are widely found in such material classes as metal alloys, minerals, ceramics, etc [2,3]. On the contrary, solid solutions formed between organic compounds are researched notably less often [4]. However, the interest in the formation of solid solutions between organic solids has significantly increased during the last decade, which is clearly indicated by the increase in the number of scientific publications investigating this phenomenon, mainly by testing previously accepted and expressing new hypotheses in the field of crystal engineering [5,6]. In this study, the formation of solid solutions in binary systems formed by thioxanthone and xanthone was explored. In each of the studied systems, mixtures of substances with different component ratios were crystallized, and powder X-ray diffractometry (PXRD) and construction of phase diagrams from thermal analysis (DSC) data were used to determine the solubility of substances in each other. The investigation of the xanthone–thioxanthone binary system reveals the existence of two solid solutions, each formed on the basis of the parent structures of xanthone and thioxanthone, respectively. One of these solid solutions exhibits miscibility of both molecules within a broad composition range (>0-80 mol% of xanthone). In addition, the crystalline structure of the solid solution involving thioxanthone:xanthone (75:25 mol%) is also presented.

Citation: Saršūns, K.; Bērziņš, A.; Rekis, T. Solid Solution Formation in Xanthone–Thioxanthone Binary System: Experimental Investigation. *Proceedings* **2023**, *92*, 67. https://doi.org/10.3390/proceedings2023092067

Academic Editors: Monika Mortimer, Anne Kahru, Ivo Leito, Riin Rebane and Villem Aruoja

Published: 30 November 2023

Author Contributions: Conceptualization, T.R.; methodology, T.R., A.B. and K.S.; software, T.R., A.B. and K.S.; validation, T.R.; formal analysis, K.S.; investigation, T.R. and K.S.; resources, T.R.; data curation, T.R. and K.S.; writing—original draft preparation, T.R., A.B. and K.S.; writing—review and editing, T.R., A.B. and K.S.; visualization, T.R., A.B. and K.S.; supervision, T.R.; project administration, A.B. and K.S.; funding acquisition, A.B. and K.S. All authors have read and agreed to the published version of the manuscript.

Funding: This research was funded through the Latvian Council of Science project, "Crystal Engineering of Pharmaceutical Multicomponent Phases for More Efficient Crystalline Phase Design" (Project No. lzp-2018/1-0312).

Institutional Review Board Statement: Not applicable.

Informed Consent Statement: Not applicable.

Data Availability Statement: The data presented in this study are available upon request from the corresponding author.

Acknowledgments: K.S. acknowledges the European Social Fund project "Strengthening of the Capacity of Doctoral Studies at the University of Latvia within the Framework of the New Doctoral Model", identification No. 8.2.2.0/20/I/006, and MikroTik Ltd. doctoral scholarship in the field of natural and medical sciences administrated by the University of Latvia Foundation.

Conflicts of Interest: The authors declare no conflict of interest. The funders had no role in the design of the study; in the collection, analyses, or interpretation of data; in the writing of the manuscript; or in the decision to publish the results.

References

1. Barbour, L.J.; Das, D.; Jacobs, T.; Lloyd, G.O.; Smith, V.J. *Supramolecular Chemistry: From Molecules to Nanomaterials*; John Wiley & Sons, Inc.: Hoboken, New Jersey, USA, 2012.
2. Vujovic, D.; Raubenheimer, H.G.; Nassimbeni, L.R. One-Dimensional CdII Coordination Polymers: Solid Solutions with NiII, Thermal Stabilities and Structures. *Eur. J. Inorg. Chem.* **2004**, *2004*, 2943–2949. [CrossRef]
3. Krishnan, R.; Swart, H.C. Luminescence Properties of Octahedrally and Tetrahedrally Coordinated Mo6+ in the Solid Solutions: Judd–Ofelt Investigation. *J. Phys. Chem. Solids* **2020**, *144*, 109519. [CrossRef]
4. Lusi, M. A Rough Guide to Molecular Solid Solutions: Design, Synthesis and Characterization of Mixed Crystals. *CrystEngComm* **2018**, *20*, 7042–7052. [CrossRef]
5. Lusi, M. Engineering Crystal Properties through Solid Solutions. *Cryst. Growth Des.* **2018**, *18*, 3704–3712. [CrossRef]
6. Braga, D. Crystal Engineering, Where from? Where To? *Chem. Commun.* **2003**, *3*, 2751–2754. [CrossRef] [PubMed]

proceedings

MDPI

Abstract

Toxicity of Silver–Chitosan Nanocomposites to Aquatic Species †

Mariliis Sihtmäe [1,*], Jüri Laanoja [1,2], Irina Blinova [1], Anne Kahru [1] and Kaja Kasemets [1]

1 Laboratory of Environmental Toxicology, National Institute of Chemical Physics and Biophysics, 12618 Tallinn, Estonia; juri.laanoja@gmail.com (J.L.); irina.blinova@kbfi.ee (I.B.); anne.kahru@kbfi.ee (A.K.); kaja.kasemets@kbfi.ee (K.K.)

2 School of Science, Tallinn Technical University, 19086 Tallinn, Estonia

* Correspondence: mariliis.sihtmae@kbfi.ee

† Presented at the International Conference EcoBalt 2023 "Chemicals & Environment", Tallinn, Estonia, 9–11 October 2023.

Keywords: antimicrobial nanomaterials; ecotoxicity; regulatory tests; bioluminescent bacteria; crustaceans

According to the World Health Organization, antimicrobial resistance (AMR) is one of the top ten global public health threats, justifying the need for new effective antimicrobials for biomedical applications. For the successful commercialization of any new material, data on its environmental safety are obligatory. Silver (Ag) compounds are widely used in wound dressings as well as disinfectants in healthcare due to their antibacterial activity and assumingly low risk of developing AMR.

In this study, the potential environmental hazard of silver–chitosan nanocomposites (nAgCSs) was evaluated. nAgCSs were synthesized by the reduction of the $AgNO_3$ with trisodium citrate and stabilized by coating with low molecular-weight chitosan (50–190 kDa). The amount of chitosan (CS) in the nanocomposites varied, resulting in three different types of nanocomposites with the weight ratio of Ag to CS 1:0.3 (nAgCS-0.3), 1:1 (nAgCS-1) and 1:3 (nAgCS-3). The toxicities of different Ag-CS nanocomposites and citrate-coated Ag nanoparticles were tested on the naturally luminescent bacterium *Vibrio fischeri* and microcrustaceans *Daphnia magna* and *Thamnocephalus platyurus*.

The primary size of nAgCSs was ~50 nm. In deionized water, the average hydrodynamic sizes were in the nanoscale (≤100 nm) and the surface charges were positive (16–26 mV). The toxicity of the studied Ag nanomaterials was evaluated using the bacterial kinetic bioluminescence inhibition and viability test [1,2] and acute immobilization/mortality tests with crustaceans [3,4]. The nAgCSs were about 10–500 fold more toxic to microcrustaceans *D. magna* (48-h EC_{50} = 0.044–0.077 mg Ag/L) and *T. platyurus* (24-h EC_{50} = 0.19–0.261 mg Ag/L) than to bacterium *V. fischeri* (30-min EC_{50}=3–26 mg Ag/L). Taking into the account the data obtained with this multi-trophic test battery the synthesized silver-chitosan nanocomposites could be classified as "extremely toxic" [L(E)C_{50} ≤ 0.1 mg/L] [5]. Interestingly, the nanocomposites with the highest chitosan content (nAgCS-3) were the most toxic to bacteria *V. fischeri* but the least toxic to the crustaceans. The data obtained show that the chitosan–silver nanocomposites may pose a hazard to aquatic organisms and must be handled accordingly.

Citation: Sihtmäe, M.; Laanoja, J.; Blinova, I.; Kahru, A.; Kasemets, K. Toxicity of Silver–Chitosan Nanocomposites to Aquatic Species. *Proceedings* **2023**, *92*, 17. https://doi.org/10.3390/proceedings2023092017

Academic Editors: Monika Mortimer, Ivo Leito, Riin Rebane and Villem Aruoja

Published: 21 November 2023

Author Contributions: Conceptualization, K.K. and A.K.; methodology, formal analysis, investigation and writing, K.K., A.K., M.S., J.L. and I.B. All authors have read and agreed to the published version of the manuscript.

Funding: This research was funded by the Estonian Research Council grant PRG 749.

Institutional Review Board Statement: Not applicable.

Informed Consent Statement: Not applicable.

Data Availability Statement: Not applicable.

Acknowledgments: This research was conducted using the NAMUR+ core facility supported by the Estonian Research Council (TT13).

Conflicts of Interest: The authors declare no conflict of interest.

References

1. *ISO 21338:2010*; Water Quality—Kinetic Determination of the Inhibitory Effects of Sediment, Other Solids and Coloured Samples on the Light Emission of *Vibrio fischeri* (Kinetic Luminescent Bacteria Test). International Organization for Standardization (ISO): Geneva, Switzerland, 2010.
2. Suppi, S.; Kasemets, K.; Ivask, A.; Künnis-Beres, K.; Sihtmäe, M.; Kurvet, I.; Aruoja, V.; Kahru, A. A novel method for comparison of biocidal properties of nanomaterials to bacteria, yeasts and algae. *J. Hazard. Mater.* **2015**, *286*, 75–84. [CrossRef] [PubMed]
3. Effects on biotic systems, test no. 202: Daphnia sp. acute immobilization test. In *OECD Guidelines for the Testing of Chemicals/Section 2*; Organization for Economic Cooperation and Development (OECD): Paris, France, 2004. [CrossRef]
4. *ISO 14380:2011*; Water Quality—Determination of the Acute Toxicity to *Thamnocephalus platyurus* (Crustacea, Anostraca). International Organization for Standardization (ISO): Geneva, Switzerland, 2011.
5. Bondarenko, O.; Juganson, K.; Ivask, A.; Kasemets, K.; Mortimer, M.; Kahru, A. Toxicity of Ag, CuO and ZnO nanoparticles to selected environmentally relevant test organisms and mammalian cells in vitro: A critical review. *Arch. Toxicol.* **2013**, *87*, 1181–1200. [CrossRef] [PubMed]

proceedings

Abstract

Silver Nanoparticles May Promote Antibiotic Resistance Gene Persistence in Wastewater Treatment Systems †

Marika Truu [1], Teele Ligi [1], Hiie Nõlvak [1], Angela Peeb [1], Kertu Tiirik [1], Arun Devarajan [1], Margit Kõiv-Vainik [1], Kuno Kasak [1], Kaja Kasemets [2] and Jaak Truu [1,*]

1 Faculty of Science and Technology, University of Tartu, 51010 Tartu, Estonia; marika.truu@ut.ee (M.T.); teele.ligi@gmail.com (T.L.); hiie.nolvak@ut.ee (H.N.); angela.peeb@gmail.com (A.P.); kertu.tiirik@gmail.com (K.T.); arun.kumar.devarajan@ut.ee (A.D.); margit.koiv.vainik@ut.ee (M.K.-V.); kuno.kasak@ut.ee (K.K.)

2 Laboratory of Environmental Toxicology, National Institute of Chemical Physics and Biophysics, 12618 Tallinn, Estonia; kaja.kasemets@kbfi.ee

* Correspondence: jaak.truu@ut.ee

† Presented at the International Conference EcoBalt 2023 "Chemicals & Environment", Tallinn, Estonia, 9–11 October 2023.

Keywords: wastewater treatment; hybrid subsurface flow filter system; silver nanoparticles; microbial community structure; silver resistance genes; antibiotic resistome

Silver nanoparticles (AgNPs) rank as some of the most commonly utilized engineered nanomaterials and are known to enter wastewater collection and treatment systems through their creation, application, and disposal processes [1]. Understanding the effects of AgNPs on the performance of wastewater treatment systems, including treatment wetlands, is critical for safeguarding public health. This research assessed how increasing levels of AgNPs and Ag^{+} ions influence the composition of microbial communities, the system treatment performance, and the elimination of antibiotic resistance genes (ARGs) in a pilot-scale hybrid wastewater treatment system. Shotgun metagenomic sequencing was employed to examine the microbial community's structure and the antibiotic resistome's diversity and composition. Quantitative PCR was used to assess the amount of ARGs. The findings indicated that heightened concentrations of AgNPs and Ag^{+} ions only moderately altered the prokaryotic community structure in the biofilms of the filter material and did not substantially affect the system's overall purification capability [2]. The addition of $AgNO_3$ caused an increase in the genetic mechanisms responsible for silver resistance in microbial communities in the vertical flow filters compared to the collargol, suggesting that the microbial communities in these biofilms can resist or adapt to the presence of silver nanoparticles. Notably, increased levels of Ag in the system did interfere with ARGs' abundance and removal efficiency, leading to an increased discharge of these genes into the environment [3]. The study also found that the concentration of accumulated Ag in the filters had a more substantial impact on the absolute and relative quantity of ARGs in the treated water than the actual amount of Ag present in the water. The research documented a higher relative presence of tetracycline, sulfonamide, and aminoglycoside resistance genes, along with increased levels of plasmid and integron-integrase genes in the biofilms of the system units treated with AgNPs. These results emphasize the need for further comprehensive studies to unravel the intricate effects of AgNPs on the nature and behavior of microbes carrying antibiotic resistance genes in wastewater treatment systems, the understanding of which is crucial for devising effective strategies to mitigate potential public health risks. Additionally, it is important to consider that the wastewater entering treatment systems contains a mixture of various engineered nanoparticles. Due to this, future research must also explore how the combination of different nanoparticles influences

Citation: Truu, M.; Ligi, T.; Nõlvak, H.; Peeb, A.; Tiirik, K.; Devarajan, A.; Kõiv-Vainik, M.; Kasak, K.; Kasemets, K.; Truu, J. Silver Nanoparticles May Promote Antibiotic Resistance Gene Persistence in Wastewater Treatment Systems. *Proceedings* **2023**, *92*, 72. https://doi.org/10.3390/proceedings2023092072

Academic Editors: Monika Mortimer, Anne Kahru, Ivo Leito, Riin Rebane and Villem Aruoja

Published: 5 December 2023

the resistome within a wastewater treatment system and the dissemination of microbes possessing antibiotic resistance genes into the environment via a system effluent.

Author Contributions: Conceptualization, J.T., M.T. and K.K. (Kaja Kasemets); methodology, J.T., M.T. and K.K. (Kaja Kasemets); investigation, M.T., T.L., H.N., K.T., M.K.-V. and K.K. (Kuno Kasak); data curation, A.P. and A.D.; writing—original draft preparation, M.T., H.N. and J.T.; writing—review and editing, M.T. and J.T.; and funding acquisition, J.T. All authors have read and agreed to the published version of the manuscript.

Funding: The work was funded by the Estonian Research Council grants PUT1125 and PRG548.

Institutional Review Board Statement: Not applicable.

Informed Consent Statement: Not applicable.

Data Availability Statement: Data sharing not applicable.

Conflicts of Interest: The authors declare no conflict of interest. The funders had no role in the design of the study, in the collection, analyses, or interpretation of data, in the writing of the manuscript, or in the decision to publish the results.

References

1. Sun, C.; Hu, K.; Mu, D.; Wang, Z.; Yu, X. The widespread use of nanomaterials: The Effects on the function and diversity of environmental microbial communities. *Microorganisms* **2022**, *10*, 2080. [CrossRef] [PubMed]
2. Truu, M.; Ligi, T.; Nõlvak, H.; Peeb, A.; Tiirik, K.; Devarajan, A.K.; Oopkaup, K.; Kasemets, K.; Kõiv-Vainik, M.; Kasak, K.; et al. Impact of synthetic silver nanoparticles on the biofilm microbial communities and wastewater treatment efficiency in experimental hybrid filter system treating municipal wastewater. *J. Hazard. Mater.* **2022**, *440*, 129721. [CrossRef] [PubMed]
3. Nõlvak, H.; Truu, M.; Tiirik, K.; Devarajan, A.K.; Peeb, A.; Truu, J. The effect of synthetic silver nanoparticles on the antibiotic resistome and the removal efficiency of antibiotic resistance genes in a hybrid filter system treating municipal wastewater. *Water Res.* **2023**, *237*, 119986. [CrossRef] [PubMed]

proceedings

MDPI

Abstract

Biocompatibility of Metal–Phenolic Network-Coated Nanoparticles †

Anett Välimets [1,2,*], Kairi Koort [2] and Monika Mortimer [1,*]

1 Laboratory of Environmental Toxicology, National Institute of Chemical Physics and Biophysics, Akadeemia tee 23, 12618 Tallinn, Estonia
2 School of Natural Sciences and Health, Tallinn University, Narva mnt. 25, 10120 Tallinn, Estonia; kairi.koort@tlu.ee
* Correspondence: anettvalimets@gmail.com (A.V.); monika.mortimer@kbfi.ee (M.M.)
† Presented at the International Conference EcoBalt 2023 "Chemicals & Environment", Tallinn, Estonia, 9–11 October 2023.

Keywords: metal–phenolic network (MPN); heavy metal; toxicity; protozoa; nanoparticles; remediation; environment

Citation: Välimets, A.; Koort, K.; Mortimer, M. Biocompatibility of Metal–Phenolic Network-Coated Nanoparticles. *Proceedings* **2023**, *92*, 33. https://doi.org/10.3390/proceedings2023092033

Academic Editors: Anne Kahru, Ivo Leito, Riin Rebane and Villem Aruoja

Published: 22 November 2023

Metal–phenolic networks (MPNs) are novel adsorbent materials that have promising applications in the environmental remediation of organic pollutants and heavy metals. For efficient adsorption, the specific surface area of MPN materials can be increased by coating nanoparticle surfaces with MPNs. Such MPN-coated nanoparticles may prove very efficient in various environmental applications; however, their safety needs to be tested at the early stages of material development. Here, free-living freshwater-ciliated protozoa were used as model organisms for testing the biocompatibility of iron–tannic acid network-coated Au nanoparticles (Fe–TA@Au NPs). Viability after 24 h Fe–TA@Au NP exposure was measured using an ATP assay kit, and intracellular reactive oxygen species (ROS) were quantified using 2′,7′-dichlorodihydrofluorescein diacetate (H_2DCFDA) [1]. Microscopy was used to qualitatively characterize the swimming behavior of protozoa as well as the uptake and depuration of Fe–TA@Au NPs in the ciliates. The results showed that Fe–TA@Au NPs were not lethal to the protozoa at the maximum concentration tested (~10^{10} particles/mL). Despite MPN-coated NP phagocytosis by protozoa and accumulation in food vacuoles, the MPNs did not affect the swimming behavior or viability during 24-h exposure. To test the performance of the novel materials as heavy metal toxicity-mitigating agents, co-exposures of protozoa to Fe–TA@Au NPs and toxic levels of copper salt ($CuSO_4$, LC_{50}~3 mg/L) were conducted. Fe–TA@Au NPs completely rescued the protozoa from cell death induced by $CuSO_4$. The underlying mechanism of toxicity mitigation could have been the removal of Cu ions by MPNs or the tannic acid in the MPNs acting as an antioxidant, as evidenced by reduced levels of ROS in the protozoa. Thus, the study showed that the effective levels of Fe–TA@Au NPs were biocompatible with unicellular freshwater model organisms and have potential for applications in metal contamination remediation in aqueous environments.

Author Contributions: Conceptualization, M.M.; methodology, A.V.; investigation, A.V.; writing—original draft preparation, A.V.; writing—review and editing, M.M. and K.K.; project administration, M.M.; funding acquisition, M.M. All authors have read and agreed to the published version of the manuscript.

Funding: This research was funded by the Estonian Research Council, grant number STP28.

Institutional Review Board Statement: Not applicable.

Informed Consent Statement: Not applicable.

Data Availability Statement: Data supporting reported results will be made available upon request.

Acknowledgments: The authors thank Heiki Vija for their assistance with chemical analysis and Maarja Otsus for their assistance with microscopy.

Conflicts of Interest: The authors declare no conflict of interest.

Reference

1. Juganson, K.; Mortimer, M.; Ivask, A.; Pucciarelli, S.; Miceli, C.; Orupõld, K.; Kahru, A. Mechanisms of toxic action of silver nanoparticles in the protozoan Tetrahymena thermophila: From gene expression to phenotypic events. *Environ. Pollut.* **2017**, *225*, 481–489. [CrossRef]

proceedings

Abstract

Mitigation of Metal Oxide Nanotoxicity with Functional Fibrils †

Yue Wang [1,2], Xiufang Liang [1,2], Fei He [3], Huayuan Tang [4], Xiang Yin [3], Nicholas Andrikopoulos [1,5], Yuhuan Li [5,6], Monika Mortimer [7], Guotao Peng [3,*] and Pu Chun Ke [1,5,*]

1 Nanomedicine Center, The Great Bay Area National Institute for Nanotechnology Innovation, 136 Kaiyuan Avenue, Guangzhou 510700, China; 202211091154@mail.scut.edu.cn (Y.W.); 202211091219@mail.scut.edu.cn (X.L.); nicholas.andrikopoulos@monash.edu (N.A.)
2 School of Biomedical Sciences and Engineering, Guangzhou International Campus, South China University of Technology, Guangzhou 510006, China
3 College of Environmental Science and Engineering, Tongji University, 1239 Siping Road, Shanghai 200092, China; 2310149@tongji.edu.cn (F.H.); 2130516@tongji.edu.cn (X.Y.)
4 College of Mechanics and Materials, Hohai University, Nanjing 211100, China; huayunt@clemson.edu
5 Drug Delivery, Disposition and Dynamics, Monash Institute of Pharmaceutical Sciences, Monash University, 381 Royal Parade, Parkville, VIC 3052, Australia; li.yuhuan1@zs-hospital.sh.cn
6 Liver Cancer Institute, Zhongshan Hospital, Key Laboratory of Carcinogenesis and Cancer Invasion, Ministry of Education, Fudan University, Shanghai 200032, China
7 Institute of Environmental and Health Sciences, College of Quality and Safety Engineering, China Jiliang University, Hangzhou 310018, China; monika.mortimer@kbfi.ee
* Correspondence: guotaopeng@tongji.edu.cn (G.P.); pu-chun.ke@monash.edu (P.C.K.)
† Presented at the International Conference EcoBalt 2023 "Chemicals & Environment", Tallinn, Estonia, 9–11 October 2023.

Keywords: amyloid fibril; metal ion; nanoparticle; binding; toxicity

Citation: Wang, Y.; Liang, X.; He, F.; Tang, H.; Yin, X.; Andrikopoulos, N.; Li, Y.; Mortimer, M.; Peng, G.; Ke, P.C. Mitigation of Metal Oxide Nanotoxicity with Functional Fibrils. *Proceedings* **2023**, *92*, 5. https://doi.org/10.3390/proceedings2023092005

Academic Editors: Anne Kahru, Ivo Leito, Riin Rebane and Villem Aruoja

Published: 21 November 2023

The toxicity of metal oxide nanoparticles has been a central research topic over the past two decades, owing to the domestic and industrial applications of this vast class of nanomaterials [1]. In the literature, ion release has been implicated as a primary cause for metal oxide nanotoxicity, coupled with the distinct physicochemical properties (e.g., large surface area, ready diffusion and dissolution, and strong adsorption) of nanoparticles, in comparison with bulk materials [2,3]. However, few solutions have been proposed thus far for overcoming the toxicity of metal oxide nanoparticles in vitro and in vivo. In this study, we engineered functional amyloid fibrils [4] using beta lactoglobulin (blg), a major whey protein, and demonstrated a scheme of ion sequestration by blg amyloid fibrils co-incubated with CuO or ZnO nanoparticles, using inductively coupled plasma mass spectrometry (ICP-MS). Our computer modeling revealed that blg fibrils possessed multiple binding sites for Cu^{2+} and Zn^{2+}, while strong binding of the metal ions often occurred at the Cys-121 residues of the fibrils. In addition, our cell viability and reactive oxygen species assays implicated blg amyloid fibrils as a functional nanomaterial with minimal toxicity. This study offered a facile engineering strategy for remediating the toxicity of metal oxide nanoparticles for facilitating their safe biological and environmental applications.

Author Contributions: Conceptualization, P.C.K.; methodology, Y.W., G.P. and X.L.; simulation, H.T.; data collection and analyses, Y.W., X.L., F.H., H.T., X.Y., N.A., Y.L., M.M. and G.P.; writing, Y.W., N.A., F.H., G.P., M.M. and P.C.K.; supervision, G.P. and P.C.K.; funding acquisition, G.P. and P.C.K. All authors have read and agreed to the published version of the manuscript.

Funding: This research was funded by the National Key Research and Development Program, Ministry of Science and Technology of China (2021YFA12009000, 2022YFC2409700), National Natural Science Foundation of China (T2250710182), and Fundamental Research Funds for the Central Universities of China.

Institutional Review Board Statement: Not applicable.

Informed Consent Statement: Not applicable.

Data Availability Statement: The data from this study may be requested to Prof. P.C. Ke.

Conflicts of Interest: The authors declare no conflict of interest.

References

1. Nel, A.; Xia, T.; Madler, L.; Li, N. Toxic potential of materials at the nanolevel. *Science* **2006**, *311*, 622–627. [CrossRef] [PubMed]
2. Ivask, A.; Juganson, K.; Bondarenko, O.; Mortimer, M.; Aruoja, V.; Kasemets, K.; Blinova, I.; Heinlaan, M.; Slaveykova, V.; Kahru, A. Mechanisms of toxic action of Ag, ZnO and CuO nanoparticles to selected ecotoxicological test organisms and mammalian cells in vitro: A comparative review. *Nanotoxicology* **2014**, *8*, 57–71. [CrossRef] [PubMed]
3. Lin, S.; Mortimer, M.; Chen, R.; Kakinen, A.; Riviere, J.E.; Davis, T.P.; Ding, F.; Ke, P.C. NanoEHS beyond toxicity—Focusing on biocorona. *Environ. Sci. Nano* **2017**, *4*, 1433–1454. [CrossRef] [PubMed]
4. Ke, P.C.; Zhou, R.; Serpell, L.C.; Riek, R.; Knowles, T.P.; Lashuel, H.A.; Gazit, E.; Hamley, I.W.; Davis, T.P.; Fändrich, M. Half a century of amyloids: Past, present and future. *Chem. Soc. Rev.* **2020**, *49*, 5473–5509. [CrossRef] [PubMed]

proceedings

MDPI

Abstract

The Optimization and Upscaling of Non-Thermal Atmospheric Plasma for Food Processing †

Yijiao Yao [1,2,*], Uta Schnabel [1], Jörg Ehlbeck [1] and Kimon-Andreas Karatzas [2]

1 Leibniz Institute for Plasma Science and Technology (INP), 17489 Greifswald, Germany; uta.schnabel@inp-greifswald.de (U.S.); ehlbeck@inp-greifswald.de (J.E.)
2 Department of Food & Nutritional Sciences, University of Reading, Whiteknights, Reading RG6 6AD, UK; k.karatzas@reading.ac.uk
* Correspondence: yijiao.yao@inp-greifswald.de; Tel.: +49-3834-554-3870
† Presented at the International Conference EcoBalt 2023 "Chemicals & Environment", Tallinn, Estonia, 9–11 October 2023.

Keywords: non-thermal plasma; FTIR spectroscopy; reactive oxygen and nitrogen species; plasma-processed air; microbial decontamination; food safety

Non-thermal plasma technology is recognized for its strong antimicrobial efficacy on food and food production environment-associated microorganisms and its ability to enhance food safety and food shelf-life [1]. However, the lack of scientific knowledge regarding the differences in physical and chemical processes at various upscaling levels hinders the successful transition of this technology from laboratory settings to industrial applications [2]. Hence, the aim of this study is to study the association between plasma-processed air (PPA) composition and its antimicrobial effects, as well as the underlying mechanism of action. This research focuses on three different air plasma torch devices, each representing a distinct stage of upscaling. A microwave plasma torch operated with compressed air delivers PPA as an antimicrobial-acting process gas. Three air plasma torch devices were applied and optimized to study the association between PPA composition and its antimicrobial effects. By employing advanced analytical methods like Fourier transform infrared spectroscopy (FTIR), a comprehensive analysis of PPA's characteristics was carried out. Key reactive nitrogen species (RNS) in the PPA were identified via spectroscopic measurements for the lab-size device MidiPLexc. From the spectrum, nitrogen monoxide (NO), nitrogen dioxide (NO_2), and dinitrogen pentoxide (N_2O_5) were detected, where NO_2 took up more than 70 % in a quantity of the key reactive species [3]. Input power and relative humidity were found to have an impact on the species concentration. The findings of this study will provide valuable insights into the physical and chemical factors that affect the scalability of non-thermal microwave plasma for antimicrobial processes. The knowledge gained will accelerate the development and application of non-thermal atmospheric pressure plasma as a highly promising alternative for disinfection and cleaning purposes.

Citation: Yao, Y.; Schnabel, U.; Ehlbeck, J.; Karatzas, K.-A. The Optimization and Upscaling of Non-Thermal Atmospheric Plasma for Food Processing. *Proceedings* **2023**, *92*, 84. https://doi.org/10.3390/proceedings2023092084

Academic Editors: Monika Mortimer, Anne Kahru, Ivo Leito, Riin Rebane and Villem Aruoja

Published: 25 January 2024

Author Contributions: Conceptualization, Y.Y., U.S. and J.E.; methodology, Y.Y.; formal analysis, Y.Y.; investigation, Y.Y.; writing—original draft preparation, Y.Y.; writing—review and editing, U.S. and K.-A.K.; supervision, U.S., J.E., and K.-A.K.; project administration, U.S. and J.E.; funding acquisition, U.S., J.E., and K.-A.K. All authors have read and agreed to the published version of the manuscript.

Funding: This project has received funding from the European Union's Horizon 2020 research and innovation program under the Marie Skłodowska Curie grant agreement No 955431.

Institutional Review Board Statement: Not applicable.

Informed Consent Statement: Not applicable.

Data Availability Statement: The data presented in this study are available on request from the corresponding author. The data are not publicly available due to ongoing work.

Conflicts of Interest: The authors declare no conflicts of interest.

References

1. Ziuzina, D.; Patil, S.; Cullen, P.J.; Keener, K.M.; Bourke, P. Atmospheric cold plasma inactivation of Escherichia coli, Salmonella enterica serovar Typhimurium and Listeria monocytogenes inoculated on fresh produce. *Food Microbiol.* **2014**, *42*, 109–116. [CrossRef]
2. Liao, X.; Liu, D.; Xiang, Q.; Ahn, J.; Chen, S.; Ye, X.; Ding, T. Inactivation mechanisms of non-thermal plasma on microbes: A review. *Food Control* **2017**, *75*, 83–91. [CrossRef]
3. Weihe, T.; Yao, Y.; Opitz, N.; Wagner, R.; Krall, J.; Schnabel, U.; Below, H.; Ehlbeck, J. Plasma-Treated Water: A Comparison with Analog Mixtures of Traceable Ingredients. *Microorganisms* **2023**, *11*, 932. [CrossRef] [PubMed]

proceedings MDPI

Abstract

Low-Temperature Synthesis and Characterization of Iron Whitlockite ($Ca_{18}Fe_2(HPO_4)_2(PO_4)_{12}$) †

Aleksej Zarkov [1,*], Diana Griesiute [1], Arita Dubnika [2,3], Dominika Zakutna [4] and Vaclav Tyrpekl [4]

1 Institute of Chemistry, Vilnius University, Naugarduko 24, LT-03225 Vilnius, Lithuania
2 Rudolfs Cimdins Riga Biomaterials Innovations and Development Centre of RTU, Faculty of Materials Science and Applied Chemistry, Institute of General Chemical Engineering, Riga Technical University, LV-1007 Riga, Latvia
3 Baltic Biomaterials Centre of Excellence, Headquarters at Riga Technical University, LV-1658 Riga, Latvia
4 Department of Inorganic Chemistry, Faculty of Science, Charles University, Hlavova 2030/8, 128 43 Prague, Czech Republic; dominika.zakutna@natur.cuni.cz (D.Z.); vaclav.tyrpekl@natur.cuni.cz (V.T.)
* Correspondence: aleksej.zarkov@chf.vu.lt
† Presented at the International Conference EcoBalt 2023 "Chemicals & Environment", Tallinn, Estonia, 9–11 October 2023.

Keywords: whitlockite; dissolution–precipitation; low temperature

Calcium phosphates (CPs) represent the most widespread class of bioceramic materials used for bone regeneration purposes due to their excellent biological performance and similarity to the natural bone. Synthetic CPs substituted with other biologically active ions can be considered as a sub-group of the CP family while possessing specific biological or physical properties provided by incorporated foreign ions. Magnesium whitlockite (Mg-WH, $Ca_{18}Mg_2(HPO_4)_2(PO_4)_{12}$) is a Mg-substituted CP, which naturally occurs in humans. This material is assumed to be the second most abundant biomineral in the human body. Although the ionic radius of Mg^{2+} is very similar to those of the first-row transition-metal ions (Zn^{2+}, Mn^{2+}, Fe^{2+}, Cu^{2+}), reports on the synthesis of such materials are very rare [1–3]. At the same time, transition-metal ions can provide some very specific properties (e.g., magnetic properties). To the best of our knowledge, there is only one work describing the synthesis and characterization of iron whitlockite (Fe-WH, $Ca_{18}Fe_2(HPO_4)_2(PO_4)_{12}$) [4]. The authors prepared Fe-WH by treating $Ca_9Fe(PO_4)_7$ with H_2 or D_2 at elevated temperatures. In our work, Fe-WH was synthesized at low temperature by the dissolution–precipitation method in aqueous medium under hydrothermal conditions. Phase conversion from brushite to Fe-WH took place in slightly acidic medium in the presence of Fe^{2+} ions just in 1 h. The crystal structure of synthesized products was confirmed by XRD analysis, FTIR, Raman and Mössbauer spectroscopies. The magnetic ordering and oxidation state information were determined using magnetic susceptibility. Morphological features were studied by means of SEM analysis. Cytotoxicity experiments were performed with MC3T3-E1 cells.

Citation: Zarkov, A.; Griesiute, D.; Dubnika, A.; Zakutna, D.; Tyrpekl, V. Low-Temperature Synthesis and Characterization of Iron Whitlockite ($Ca_{18}Fe_2(HPO_4)_2(PO_4)_{12}$). *Proceedings* **2023**, *92*, 25. https://doi.org/10.3390/proceedings2023092025

Academic Editors: Monika Mortimer, Anne Kahru, Ivo Leito, Riin Rebane and Villem Aruoja

Published: 21 November 2023

Author Contributions: Conceptualization, A.Z. and D.G.; methodology, A.Z. and D.G.; formal analysis, D.G., A.D., D.Z. and V.T.; investigation, D.G., A.D., D.Z. and V.T.; writing—original draft preparation, A.Z.; writing—review and editing, A.Z.; visualization, D.G.; supervision, A.Z.; funding acquisition, A.Z. All authors have read and agreed to the published version of the manuscript.

Funding: This work was funded by the grant WHITCERAM (No. S-LJB-22-1) from the Research Council of Lithuania. A.D. acknowledge the access to the infrastructure and expertise of the BBCE H2020 GA857287.

Institutional Review Board Statement: Not applicable.

Informed Consent Statement: Not applicable.

Data Availability Statement: Not applicable.

Conflicts of Interest: The authors declare no conflict of interest.

References

1. Kizalaite, A.; Grigoraviciute-Puroniene, I.; Asuigui, D.R.C.; Stoll, S.L.; Cho, S.H.; Sekino, T.; Kareiva, A.; Zarkov, A. Dissolution–Precipitation Synthesis and Characterization of Zinc Whitlockite with Variable Metal Content. *ACS Biomater. Sci. Eng.* **2021**, *7*, 3586–3593. [CrossRef]
2. Kizalaite, A.; Klimavicius, V.; Versockiene, J.; Lastauskiene, E.; Murauskas, T.; Skaudzius, R.; Yokoi, T.; Kawashita, M.; Goto, T.; Sekino, T.; et al. Peculiarities of the formation, structural and morphological properties of zinc whitlockite ($Ca_{18}Zn_2(HPO_4)_2(PO_4)_{12}$) synthesized via a phase transformation process under hydrothermal conditions. *Crystengcomm* **2022**, *24*, 5068–5079. [CrossRef]
3. Heng, C.; Liu, W.; Zheng, X.; Ma, X.; Hui, J.; Fan, D. Dopamine and DNA functionalized manganese whitlockite nanocrystals for magnetic resonance imaging and chemo-photothermal therapy of tumors. *Colloids Surf. B Biointerfaces* **2023**, *222*, 113120. [CrossRef]
4. Belik, A.A.; Izumi, F.; Stefanovich, S.Y.; Lazoryak, B.I.; Oikawa, K. Chemical and Structural Properties of a Whitlockite-like Phosphate, $Ca_9FeD(PO_4)_7$. *Chem. Mater.* **2002**, *14*, 3937–3945. [CrossRef]

proceedings

MDPI

Abstract

Pilot Study of Grassland Soil Soluble Organic Matter with High-Resolution Mass Spectrometry †

Anna Zobel

Institute of Chemistry, Faculty of Science and Technology, University of Tartu, 50411 Tartu, Estonia; anna.zobel5@gmail.com

† Presented at the International Conference EcoBalt 2023 "Chemicals & Environment", Tallinn, Estonia, 9–11 October 2023.

Keywords: soil; soil health; soil organic matter; high-resolution mass spectrometry; FT-ICR MS

Soil is the largest terrestrial carbon pool and regulates carbon, water, and nutrient cycles [1,2]. The diversity of the soil environment prevents it from having one certain parameter to assess the condition of the soil. Instead, different physical, chemical, and biological parameters are used to assess soil health in general [1,3]. A non-targeted method was proposed for fingerprinting soil soluble organic matter (SOM) with high-resolution mass spectrometry (HRMS) and making assessments of soil health from collected data. Soil samples were collected from a permanent grassland in Tõravere over the course of April 2022 to June 2022. The samples were dried with a vacuum and extracted with organic solvents. Three solvents were compared for extraction—acetonitrile, methanol, and toluene. The extracts were analyzed with HRMS by using a Fourier-transform ion cyclotron resonance mass spectrometer (FT-ICR MS). Flow infusion together with two ionization sources were compared—nano-electrospray ionization (nESI) and atmospheric pressure chemical ionization (APCI) [4]. The results demonstrated that the mass spectra of the methanol extracts were most abundant in peaks. Further data analysis of the elucidated molecular formulas revealed that the identified compounds in each extract consisted mostly of various lipids, although some peptides were identified in extracts of all three solvents and some carbohydrates were identified in the methanol and toluene extracts. The results also suggest nESI to be more suitable for HRMS analysis since it used up smaller amounts of extract but gave more peaks on the mass spectrum [4]. Further research on more diverse sample types and variations of additional parameters (time-series for observation of the effects of the plant development cycle and soil microorganisms' population dynamics, changing weather conditions, etc.) are needed for looking at relationships between SOM chemical composition and soil health.

Citation: Zobel, A. Pilot Study of Grassland Soil Soluble Organic Matter with High-Resolution Mass Spectrometry. *Proceedings* **2023**, *92*, 2. https://doi.org/10.3390/proceedings2023092002

Academic Editors: Monika Mortimer, Anne Kahru, Ivo Leito, Riin Rebane and Villem Aruoja

Published: 20 November 2023

Funding: This research was funded by the Estonian Research Council, grant number TT4.

Institutional Review Board Statement: Not applicable.

Informed Consent Statement: Not applicable.

Data Availability Statement: The data presented in this study are available on request from the corresponding author. The data are not publicly available due to the still on-going research.

Acknowledgments: I am deeply grateful to my supervisors, Ivari Kaljurand and Koit Herodes, who both made this work possible. Many thanks to my chair, Chair of Analytical Chemistry, for the assistance and support given during the research. I thank the University of Tartu's Department of Botany for giving useful insights into the topic of soil.

Conflicts of Interest: The author declares no conflict of interest.

References

1. Rinot, O.; Levy, G.J.; Steinberg, Y.; Svoray, T.; Eshel, G. Soil Health Assessment: A Critical Review of Current Methodologies and a Proposed New Approach. *Sci. Total Environ.* **2019**, *648*, 1484–1491. [CrossRef] [PubMed]
2. Communication from the Commission to the European Parliament, the Council, the European Economic and Social Committee and the Committee of the Regions EU Soil Strategy for 2030 Reaping the Benefits of Healthy Soils for People, Food, Nature and Climate. 2021. Available online: https://eur-lex.europa.eu/legalcontent/EN/TXT/?uri=CELEX:52021DC0699 (accessed on 20 May 2023).
3. Bünemann, E.K.; Bongiorno, G.; Bai, Z.; Creamer, R.E.; De Deyn, G.; de Goede, R.; Fleskens, L.; Geissen, V.; Kuyper, T.W.; Mäder, P.; et al. Soil Quality—A Critical Review. *Soil Biol. Biochem.* **2018**, *120*, 105–125. [CrossRef]
4. Zobel, A. Pilot Study of Grassland Soil Soluble Organic Matter with High Resolution Mass Spectrometry. Bachelor's Thesis, University of Tartu, Tartu, Estonia, 2023.

MDPI AG
Grosspeteranlage 5
4052 Basel
Switzerland
Tel.: +41 61 683 77 34

Proceedings Editorial Office
E-mail: proceedings@mdpi.com
www.mdpi.com/journal/proceedings

www.ingramcontent.com/pod-product-compliance
Lightning Source LLC
Chambersburg PA
CBHW061604210926
51850CB00009B/2211
* 9 7 8 3 7 2 5 8 1 5 5 7 9 *